TECHNOLOGY AND INNOVATION IN JAPAN

Japan has advanced to a leading position in numerous fields of technology since the beginning of the 1980s. The Japanese have also been major suppliers of technology to the rest of Asia. However, as Japan's technological level has risen and technology trade with foreign countries has become more balanced, acquiring leading-edge technology from external sources has become increasingly difficult. In order to preserve long-term international competitiveness, Japan must strengthen its domestic system of innovation.

Technology and Innovation in Japan demonstrates that long-awaited changes to technology policy and corporate strategy are now taking place. This study also questions whether it is the program of reform or the will and ability to implement these reforms which is new. Contributors explore the repositioning of the Japanese science and technology system on three levels:

- institutional structure and technology policy;
- organizational and managerial changes and business–government relations;
- developments in key technology sectors.

Hemmert and Oberländer have brought together Japanese, European and American contributors from fields of research as varied as business studies, medicine, economics, physics and policy sciences. From these diverse perspectives, all arrive at a single conclusion: at the threshold of the twenty-first century Japanese policy-makers and managers have become serious about science and technology reform.

Martin Hemmert is a Research Fellow with the German Research Association (DFG) at Essen University, Germany.
Christian Oberländer is Research Associate at the Department of Surgery of The University of Tokyo, Japan.

ROUTLEDGE STUDIES IN THE GROWTH ECONOMIES OF ASIA

1 THE CHANGING CAPITAL MARKETS OF EAST ASIA
Edited by Ky Cao

2 FINANCIAL REFORM IN CHINA
Edited by On Kit Tam

3 WOMEN AND INDUSTRIALIZATION IN ASIA
Edited by Susan Horton

4 JAPAN'S TRADE POLICY
Action or reaction?
Yumiko Mikanagi

5 THE JAPANESE ELECTION SYSTEM
Three analytical perspectives
Junichiro Wada

6 THE ECONOMICS OF THE LATECOMERS
Catching-up, technology transfer and institutions in Germany, Japan and South Korea
Jang-Sup Shin

7 INDUSTRIALIZATION IN MALAYSIA
Import substitution and infant industry performance
Rokiah Alavi

8 ECONOMIC DEVELOPMENT IN TWENTIETH-CENTURY EAST ASIA
The international context
Edited by Aiko Ikeo

9 THE POLITICS OF ECONOMIC DEVELOPMENT IN INDONESIA
Contending perspectives
Edited by Ian Chalmers and Vedi Hadiz

10 STUDIES IN THE ECONOMIC HISTORY OF THE PACIFIC RIM
Edited by Sally M. Miller, A. J. H. Latham and Dennis O. Flynn

11 WORKERS AND THE STATE IN NEW ORDER INDONESIA
Vedi R. Hadiz

12 THE JAPANESE FOREIGN EXCHANGE MARKET
Beate Reszat

13 EXCHANGE RATE POLICIES IN EMERGING ASIAN COUNTRIES
Edited by Stefan Collignon, Jean Pisani-Ferry and Yung Chul Park

14 CHINESE FIRMS AND TECHNOLOGY IN THE REFORM ERA
Yizheng Shi

15 JAPANESE VIEWS ON ECONOMIC DEVELOPMENT
Diverse paths to the market
Kenichi Ohno and Izumi Ohno

16 TECHNOLOGICAL CAPABILITIES AND EXPORT SUCCESS IN ASIA
Edited by Dieter Erust, Tom Ganiatsos and Lynn Mytelka

17 TRADE AND INVESTMENT IN CHINA
The European experience
Edited by Roger Strange, Jim Slater and Limin Wang

18 TECHNOLOGY AND INNOVATION IN JAPAN
Policy and management for the twenty-first century
Edited by Martin Hemmert and Christian Oberländer

19 TRADE POLICY ISSUES IN ASIAN DEVELOPMENT
Prema-chandra Athukorala

20 ECONOMIC INTEGRATION IN THE ASIA PACIFIC REGION
Ippei Yamazawa

TECHNOLOGY AND INNOVATION IN JAPAN

Policy and management for the twenty-first century

Edited by
Martin Hemmert and Christian Oberländer

ROUTLEDGE

London and New York

First published 1998
by Routledge
11 New Fetter Lane, London EC4P 4EE

Simultaneously published in the USA and Canada
by Routledge
29 West 35th Street, New York, NY 10001

Typeset in Garamond by RefineCatch Limited, Bungay, Suffolk
Printed and bound in Great Britain by
Biddles Ltd, Guildford and King's Lynn

British Library Cataloguing in Publication Data
A catalogue record for this book is available from the British Library

Library of Congress Cataloging in Publication Data
Technology and innovation in Japan: policy and management for the twenty-first century / [edited by] Martin Hemmert and Christian Oberländer.
Includes bibliographical references and index.
1. Technology and state – Japan. 2. Technology – Japan – Management.
I. Hemmert, Martin. II. Oberländer, Christian.
T27.J3T42 1998
338.95207 – dc21 98–13626

ISBN 0–415–19152–1

CONTENTS

FIGURES

TABLES

CONTRIBUTORS

Brendan Barker, Science Officer, The British Council, Tokyo Office.

Michael Becht, Research Associate, Superconductivity Research Laboratory, International Superconductivity Technology Center.

Akira Goto, Professor, Institute of Innovation Research and Faculty of Economics, Hitotsubashi University; Research Director, National Institute of Science and Technology Policy (NISTEP).

Martin Hemmert, Research Fellow, German Research Association (DFG) at Essen University.

Tomohiro Ijichi, Research Associate, National Institute of Science and Technology Policy (NISTEP).

Koki Inoue, Senior Researcher, Economic Research Institute, Japan Society for the Promotion of Machine Industry (KSK).

Ken Kusunoki, Associate Professor, Faculty of Commerce and Institute of Innovation Research, Hitotsubashi University.

Christian Oberländer, Research Associate, Department of Surgery, The University of Tokyo.

Hiroyuki Odagiri, Professor, Graduate School of Economics, Hitotsubashi University.

Tim Ray, Research Fellow, Policy Research in Engineering, Science and Technology (PREST), University of Manchester.

Tanya Sienko, Researcher, Corporate Development Laboratory, Sumitomo 3M Ltd.

Chihiro Watanabe, Professor, Department of Industrial Engineering and Management, Tokyo Institute of Technology.

Shinichi Yamamoto, Professor, Research Center for University Studies, University of Tsukuba.

FOREWORD

After the Second World War, Japan's economy underwent a period of rapid economic growth and has now drawn equal with other industrialized countries in terms of productivity. A crucial driving force behind this growth process was technological progress: Japan managed to close the technological gap.

In the first decades after the war, technology transfer from the US and Western Europe played a major role in advancing Japan's technological level. During this period, the Japanese system of science and technology was oriented towards the quick distribution and utilization of technological know-how. However, since the beginning of the 1980s, Japan has advanced to a position of leadership in numerous fields of technology. In order to preserve her long-term international competitiveness, the strengthening of Japan's domestic system of technology generation has become a crucial task.

In the 1990s, the need for structural change in the technology and innovation system is generally acknowledged in Japan. There is a continuous discourse, however, about what strategies and measures should be actually taken to facilitate the transformation process. From a Western perspective, the course of this discussion and the strategies and policies resulting from it are important also. As Japan is now a major economic and technological power, the innovation system of this country has become a significant factor for international competition and global economic development. Moreover, since technological interdependence on the international level is rapidly progressing, mutual understanding of the institutional setting in other countries is of growing importance as a prerequisite for smooth and successful international cooperation in the field of science and technology.

In June and July 1996, a series of international workshops were held at the German Institute for Japanese Studies in Tokyo about the recent structural change in Japan's system of technology and innovation. Japanese and Western experts discussed in detail the recent state of the ongoing debate in Japan. The Japanese innovation system was analyzed on three levels: (1) institutional structure; (2) main functions of technology policy, technology management and technology development; and (3) implementation in selected key technology sectors.

We would like to take this opportunity to express our thanks to all participants of the workshop series who contributed to its success through their interesting presentations and their lively discussion of the wide range of issues that were covered. Martin Hemmert, Robert Horres, Edzard Janssen, and Christian Oberländer from the German Institute for Japanese Studies were in charge of planning and organization. The workshops and the publication were jointly supported by the Friedrich-Ebert-Foundation and the German Institute for Japanese Studies. Finally, we would like to thank Routledge for its efforts to ensure the swift publication of this book.

Dr Michael Ehrke
Friedrich-Ebert-Foundation

Professor Dr Josef Kreiner
Former Director, German Institute for Japanese Studies
Philipp-Franz-von-Siebold Foundation

ACKNOWLEDGEMENTS

First of all, we would like to express our thanks to all contributors of this volume, which is based on a research project that brought together people from various countries and regions (Europe, the United States, Japan, etc.) and with very diverse academic and professional backgrounds. While we firmly believe that intercultural and interdisciplinary work is a fruitful approach, we are also aware that the project demanded a considerable amount of flexibility and patience from many of the contributors

Special thanks go to Professor Akira Goto of Hitotsubashi University and the National Institute for Science and Technology Policy (NISTEP) who gave us substantial advice on the concept of the workshop series this book is based on, and also readily provided his expertise in discussing recent issues concerning the Japanese system of technology and innovation.

The workshop series as well as the preparation of this book were financially supported by the Friedrich-Ebert-Foundation, Tokyo Office, and the German Institute for Japanese Studies (DIJ).

Daniel Dirks and Hendrik Meyer-Ohle (DIJ), Robert Horres (Bonn University), Franz Waldenberger (Munich University), and Christian Schönbach of Chūgai's Institute for Molecular Medicine assisted us in the review process. Keiko Asano and Michaela Lincke of DIJ provided great support in organizing the workshops. Finally, we wish to thank Claus Harmer (DIJ), who provided technical assistance in software issues and Phillip Anderson, who took charge of the English language editing.

Martin Hemmert
Christian Oberländer
Tokyo, December 1997

Part I

INTRODUCTION

1

THE JAPANESE SYSTEM OF TECHNOLOGY AND INNOVATION

Preparing for the twenty-first century

Martin Hemmert and Christian Oberländer

Fundamental change, gradual evolution, or stagnation?

Since the beginning of the 1980s, Japan has advanced to a leading position in numerous fields of high technology. This is reflected, for example, in Japan's technology trade (Figure 1.1). From 1975 until 1984, exports of technology continually approached imports. After 1985, the expansion of exports temporarily slowed down, apparently caused by the appreciation of the Yen after the Plaza accord. In 1993, finally, exports exceeded imports for the first time. In particular, Japan has reduced the technology trade gap with the US and Europe significantly, raising the ratio of exports to imports from 0.13 and

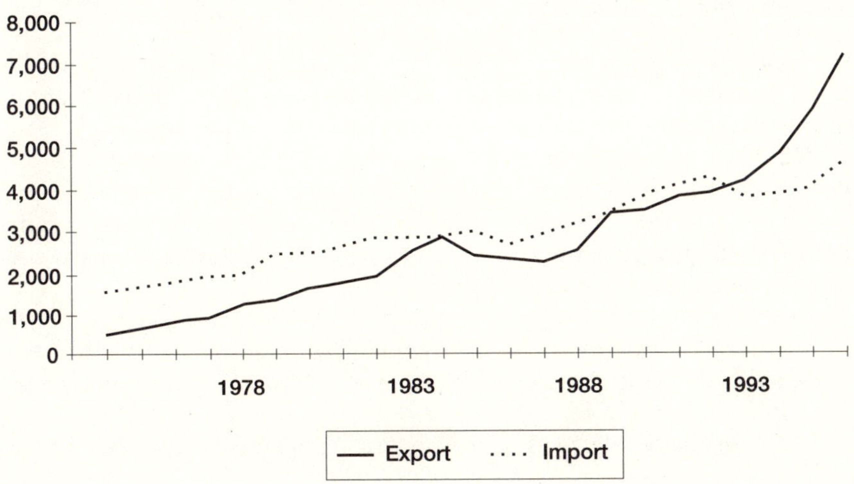

Figure 1.1 Japan's technology exports and imports, 1975–96 (100 million Yen)
Source: Somucho Tokeikyoku (1997 and various years).

0.23 in 1975 to 0.58 and 0.86 in 1995 respectively. At the same time, Japan has built a strong position as a technology supplier to Asia, with exports to that region making up 50.2 percent of total technology exports by 1995 (Somucho Tokeikyoku 1997: 28).

While Japan's technological level has risen and technology trade with foreign countries has become more balanced, acquiring leading edge technology from foreign countries is increasingly difficult. In order to preserve its long-term international competitiveness, the strengthening of Japan's domestic system of innovation has become crucial. On the surface, this necessity has been emphasized by Japanese policy-makers since at least the 1960s. For a long time, little did in fact happen to enable the Japanese innovation system to restructure itself. Recently, however, several indicators signal a change in Japan's science and technology policy as well as in corporate strategy, responding to this challenge (Odagiri and Goto 1993: 109–10). It appears likely that policy-makers and managers at the threshold of the twenty-first century have become serious about their proclamations that there is a need for increased funding and organizational reform in science and technology.

However, these developments have not yet been scrutinized sufficiently, even in the most recent survey literature. Goto and Odagiri (1997), for example, while providing a thorough overview of numerous facets of the Japanese national innovation system, focus their discussion on the 1970s and 1980s. Furthermore, most of the contributors to their volume choose an economic approach, providing broad outlines of certain elements of the innovation system rather than detailed case studies. In contrast, this volume tries to provide some answers to the many open questions about the developments in the 1990s using perspectives as diverse as possible. Are the ensuing policy changes fundamental in character? Or are they part of a slow but ongoing evolution? Are there forces resisting change? The papers in this book focus on the period from 1985 to 1995 and give hints on what the twenty-first century might have in stock for the Japanese innovation system. The approaches taken to analyze the present situation of the Japanese science and technology system are as numerous as the fields from which the contributors come: business studies, economics, medicine, physics, policy sciences, etc. In particular, they examine to what extent a change toward domestically achieved technological advances is taking place. The Japanese science and technology system is scrutinized on three levels:

1 institutional structure and technology policy;
2 organizational and managerial changes and business–government relations; and
3 developments in selected key technology sectors concerning one or both of the former areas.

The following sections discuss the papers' findings under various topical facets, adding some of our own comments regarding certain recent developments in the Japanese science and technology system.

Institutional structure and policy process

The enactment of the Science and Technology Basic Law (*Kagaku Gijutsu Kihon-hō*) on 15 November 1995 is treated by many as the most recent pivotal event in Japanese science policy (for example Plate 1996). The Basic Law requires the Japanese administration to raise science- and technology-related spending for five consecutive fiscal years. Moreover, the law provides for the drafting of the Science and Technology Basic Plan (*Kagaku Gijutsu Kihon Keikaku*) by the government. This Plan, which the cabinet approved on 2 June 1996, aims at doubling the science and technology budget by the year 2000, as compared with 1992.

The way in which this law was enacted, by a multiparty group of diet members under the leadership of a former Ministry of International Trade and Industry (MITI) bureaucrat rather than by the government, signals that this policy measure enjoys broad support, as it was proposed by politicians with very diverse backgrounds and not simply by a single ministry, which is normally the case in Japanese legislation. This broad support is strongly rooted in the belief that Japan can no longer continue 'catching up' and that it is now a forerunner in science and technology. In order to withstand international economic competition, to 'sustain high industrialized standards' and to increase its international contribution, an expansion of Japan's R&D efforts is viewed as essential (NSF Tokyo Office 1996). To achieve these ends, government sponsored basic research especially should be expanded. The Plan lists the following measures to be taken (Kagaku Gijutsu Kaigi 1996):

- government aid, particularly for the work of young researchers;
- more joint industry–university–government research;
- more international exchange;
- expanding R&D spending.

A comparison of the new Basic Law and Plan with its predecessors, the 11th and 18th recommendations of the Council for Science and Technology (CST, *Kagaku Gijutsu Kaigi*) of 1984 and 1992 which formed the basis of the General Guidelines for Science and Technology Policy of 1986 and 1992, raises some interesting points. The recommendations of 1984 already called for greater government spending for basic research, incentives for industry to invest heavily in that area, and greater cooperation between academia, industry and national laboratories. Besides this, the creativity of young researchers, the harmonization of R&D with mankind and society, and the internationalization of Japanese R&D were to be furthered (*Japan Economic Journal*, 4

December 1984). The 18th recommendation of 1992 essentially reiterated the same points (Horres and Kreiner 1992: 14). Thus, there is little of an 'original' program in the new Basic Law and Plan. This hints at considerable continuity in Japan's R&D policy and casts doubt on suggestions of a major 'policy adjustment' (as in Plate 1996: 659).

Both the recommendations of 1984 and the Basic Law and Plan of 1995/96 can be interpreted as reactions to the decreasing role of government in research spending (Figure 1.2). This is mainly a phenomenon of the 1980s, a period when strict fiscal discipline was enforced across all sectors and government thus reduced its share in Japan's overall R&D spending. This led to a corresponding rise in the share of the R&D burden borne by industry. A similar development occurred concerning the performing sectors of R&D. As can be seen from Figure 1.3, the share of universities and research institutions as performers of R&D decreased significantly between 1985 and 1990, from 20.1 percent to 17.6 percent, and from 13.1 percent to 11.6 percent, respectively.

However, this development was reversed in the 1990s, when the bursting of the 'economic bubble' put pressure on corporate R&D spending. At the same time, the Japanese government attempted to stimulate economic activity through fiscal measures, thus providing a climate favorable for the expansion of public R&D spending. This allowed for a temporary recovery of the share of government in overall R&D expenditures. In parallel, the share of universities and research institutions as R&D performers recovered to 20.7 percent and 14.1 percent, respectively, in 1995.

There is a possibility, however, that the trend of a declining role for government in R&D financing will continue in the future as ever stronger pressure for fiscal discipline is mounting (Tomita 1996), and as companies recovering from the recession of 1992/93 are strengthening anew their R&D

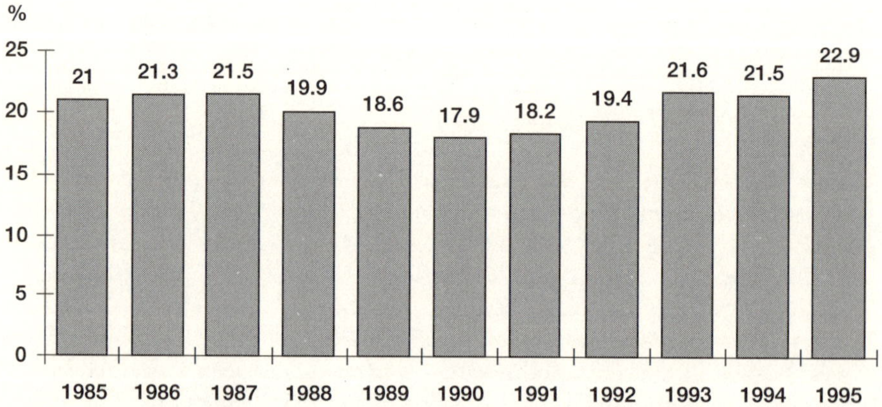

Figure 1.2 Government's share in financing Japan's R&D (percent)
Source: Somucho Tokeikyoku (1997 and various years).

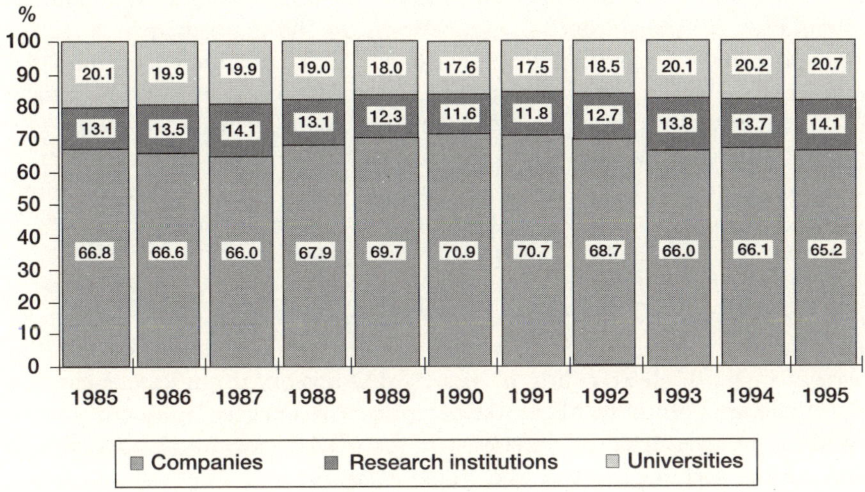

Figure 1.3 Performing sectors of R&D in Japan
Source: Somucho Tokeikyoku (1997 and various years).

efforts. In this environment, and counter to the prior recommendations, the Plan might serve to protect the science and technology budget from global spending ceilings, as they were enforced in the 1980s. If the Plan succeeds in ensuring a continued increase in government's R&D funding, its effect would far exceed that of earlier Guidelines. The fact that the Hashimoto administration refuted the resistance of the powerful Ministry of Finance (MOF), which had opposed the inclusion of numerical targets in the Plan, initially indicated that the Plan may actually work in this way (*Nikkei Weekly*, 10 June 1996). For the fiscal year 1997, expenditure for science and technology was to grow by 11.9 percent across all ministries, compared to a rise in total government spending of only 3 percent. MITI's R&D budget even more than doubled (Nathan 1997: 104). However, already one year after the approval of the Basic Plan, in the summer of 1997, budget austerity seriously threatened to put the Plan's realization in jeopardy. According to current reform plans, expenditures are to be limited to a 5 percent increase, and no new large-scale projects are to be approved until the year 2003. Some sources have even spoken of a burst of the 'research money bubble' (Triendl 1997b: 643).

When pondering the structure of the Japanese innovation system, it must be considered that government's role in R&D funding is generally very small. As can be seen from Figure 1.2, only about one-fifth of Japan's total R&D budget is supplied by the government. This share is substantially lower than in all other major industrialized countries, where government's funding proportion is somewhere between 30 percent and 45 percent

(OECD 1996: 20). Furthermore, in contrast to all other advanced economies, a significant net transfer of R&D funds from the public to the private sector does not occur in Japan (OECD 1996: 20–3, 31). These differences compared to other advanced countries are explained by the lateness of Japan's industrial development (Sakakibara 1995: 162–3). It seems that in the economic race to catch up, the Japanese government never had the time (and, until recently, never felt the necessity) to build up a publicly funded R&D infrastructure comparable in size to those in the US or in Western European countries.

Among the papers in this volume, the role of government in research funding is scrutinized in particular by Ijichi and Goto. They find overall Japanese R&D expenditure to be decreasing in 1992 for the first time in postwar Japan while spending on basic R&D bottomed out earlier and has been increasing since 1990. However, in international comparison, Japan's spending on basic R&D as a share of its total R&D expenditure is still low. Although government is still the largest source of funding for basic R&D, Japanese industry is – by international standards – an exceptionally large contributor to basic R&D in Japan, a position it acquired since the second half of the 1970s. This, too, is perhaps due to reduced government spending in the 1980s while other explanations include, for example, the weakness of Japanese university research. Ijichi and Goto trace attempts to promote basic research back to the General Guidelines of 1986 which were later supplemented by several national programs (research in prioritized areas, inter-ministry basic research, centers of excellence (COE)). Public research spending has been raised since 1992 as a fiscal stimulus measure and by regrouping it as public investment rather than general expenditure. Although the Basic Plan continues this trend, Ijichi and Goto emphasize that industry expenditure for basic R&D needs to be increased as well.

Watanabe and Hemmert give an overview of Japan's technological development since the Second World War from the perspective of the positive interaction between capital investment and technological progress. They point out that the fast technological development of Japan's manufacturing industry has to be understood in the overall context of the high-growth economy. It was supported by a favorable environment, including socioeconomic factors and technology policy. An econometric analysis reveals that R&D expenditures and capital investment complemented each other until the 'bubble economy' in the second half of the 1980s, when they became substitutes for each other. Further calculations show that the R&D expenditures of Japanese manufacturers, measured at constant prices, began to stagnate at that time, indicating that their subsequent decrease in the 1990s had their origins in a deviation of managerial attention from R&D during the 'bubble economy.' Watanabe and Hemmert suggest that reinforced R&D funding as provided by the Basic Plan may be crucial to revitalize the positive feedback loop between capital investment and technology.

Organization, management and business–government relations

Apart from funding, the Basic Plan has been criticized by the Japanese press on several acounts. It has been noted that the Plan lacks a clear 'philosophy for why the government seeks the rapid development of science and technology.' Moreover, there is concern that the Plan might give rise to a suspicion of future Japanese military adventures among Japan's neighbors (*Nikkei Weekly*, 22 July 1996). However, the solution of organizational problems appears to be the most important problem. Inter-agency policy overlap, intra-governmental competition, and excessive regulations are named. Since these areas appear crucial for the effectiveness of the Basic Plan, the next section will investigate which initiatives have been taken here.

Traditionally, Japanese R&D units, public as well as private, have been characterized by life-time employment and low inter-organizational mobility of researchers. Moreover, the level of international involvement of most Japanese R&D organizations has been relatively low. These factors, while contributing to the stability of research organizations, have been identified as major obstacles for improvement in the output performance of the Japanese innovation system (Sakakibara 1995: 255–61). In the following paragraphs, the question to what extent organizational and managerial change has occurred in the Japanese innovation system is reviewed on three levels: the intra-organizational, the inter-organizational, and the international.

The intra-organizational level

For the intra-organizational level, several reform measures have frequently been discussed regarding R&D units in the business innovation system. Odagiri's paper addresses the issue of the need for a paradigm change in the Japanese business innovation system from a theoretical perspective. After a brief review of how Japan managed to adapt and improve foreign technology in its historical catch-up process, some basic features of technological know-how as a business resource are discussed. Path-dependency of firms' abilities to invent or assimilate technologies and inter-organizational spill-over of technological know-how are emphasized as central determinants of business performance with respect to innovation. It is argued that, in the context of these factors, certain Japanese business practices like life-time employment and close vertical inter-firm relations have contributed to the strong technological performance of manufacturing firms. Therefore, the author concludes, a total reorientation of R&D management risks sacrificing the fundamental strengths which companies have acquired over decades. Rather, a partial adjustment is needed to enable Japanese companies to internationalize their innovation system smoothly.

Hemmert complements Odagiri's findings by an empirical analysis of the recent development in industrial R&D management. Based on the hypothesis that companies are under simultaneous pressure to (a) restructure R&D to strengthen their own technological base, and (b) rationalize and downsize to improve their short- and medium-term business performance, a quantitative assessment of manufacturing firms' R&D performance is conducted, and strategies in several managerial fields are scrutinized. The quantitative overview shows that R&D expenditures have been significantly reduced in the first half of the 1990s, indicating a priority for rationalization. The managerial analysis confirms that, with regard to organizational structure, a rationalization strategy is predominant, while in personnel management reform is rather limited in scope. The introduction of annual, performance-based salary systems (*nenpōsei*) and of flexible working hours (*sairyō rōdōsei*) has been announced by many large manufacturing firms. A closer look at the measures actually taken reveals, however, that they often appear to be experimental in character. Only R&D personnel management in most Japanese firms remains largely unchanged for the time being. The most dynamic area among efforts to strengthen industry's R&D still seems to be internationalization, namely the setup of R&D facilities abroad. Altogether, many industrial firms appear to pursue a mixed approach to overcome the strategic dilemma pointed out above.

In universities and public research institutions, the focus of the discussion about intra-organizational change is somewhat different. Throughout the last fifteen years, different types of new R&D organizations have been established in addition to those already in existence. The program that seems to have caught the most attention among them is ERATO (Exploratory Research for Advanced Technology) due to its highly innovative organizational approach, as Barker points out.

Kusunoki undertook a comparative empirical study of ERATO and a MITI laboratory which he views as representative for the 'dynamic-network' and the 'bureaucratic' organization, respectively. From interviews and a survey, he finds significant differences between the two types of organizations not only in the degree to which outside communication is pursued, but also in the way it is utilized and the performance it leads to. Kusunoki suspects management and organizational characteristics to be the main causes not only of diminished outside communication, but also of difficulties in applying its results within the bureaucratic organization. In the dynamic network organization, these problems appear to have been overcome; Kusunoki therefore suggests that this form of organization should be expanded. However, he also notes that both types of organization are necessary for a stable research environment.

The architecture of the new research units is more open than before and their performance appears to be substantially better than that of the older ones. A remaining problem is the restructuring of the older, bureaucratic-type R&D organizations that still account for the bulk of the public research institutions in Japan. Recently discussed measures include increased competition for research funds and the strengthening of the evaluation of R&D units. In 1997, the Science and Technology Agency (STA) developed a National Guideline on the Method of Evaluation for Government R&D (Kagaku Gijutsucho 1997). According to this detailed Guideline, all governmental R&D institutions shall be evaluated in the future based on the following principles:

- explicitness of evaluation standards and processes;
- inclusion of outsiders into the evaluations;
- public accessibility of the evaluation results;
- reflection of the evaluation results in the future distribution of R&D funds.

Based on this Guideline, a large number of governmental R&D units are scheduled for evaluation throughout the next two to three years. Since these approaches are very new, however, it is too early for an assessment of their actual impact on the performance of public research institutions. In any case, a considerable amount of resistance has to be taken into account when applying such measures. A recent survey among researchers in public laboratories, for instance, revealed that many of them are totally opposed to evaluation (*Nihon Keizai Shinbun*, 30 December 1996). Albeit slowly, pressure for stricter evaluation is, however, mounting in the context of fiscal reform. While first guidelines for evaluation were already issued in 1986, a report by a subcommittee of the prime minister's office branded the present state of evaluation as 'unsatisfactory' in May 1997. In the same year, MITI planned to create a division for research evaluation at its Agency for Industrial Science and Technology (AIST). Besides an initial peer review, all research proposals will be reviewed once during their lifetime and their duration will be limited to a maximum of five years (Triendl 1997a: 444).

Another recently discussed reorganization move that is heading in the same direction is an initiative to allow time-limited working contracts for researchers in public R&D institutions (*Nihon Keizai Shinbun*, 4 October 1997). The necessary legislative steps have already been taken, providing the opportunity for research institutions to apply two such types of contracts as alternatives to the hitherto rule of unlimited contracts: one to invite eminent senior researchers and one for young R&D workers as an opportunity for postdoctoral qualification. As the new system started in June 1997, however, the same situation as that concerning the evaluation of public R&D units applies. It is still too early to assess to what extent it will be used in practice and how large its impact will be on the organizational routines in the respective research units.

Japan's higher education is known for its high level of university enrollment and credited for its pivotal role in the development of the Japanese economy. In research, its main functions are to train researchers, perform research and development, and identify highly qualified human resources. After a brief review of the postwar development of the higher education system, Yamamoto interprets university faculties' strong focus on graduate education as a strategy to escape from the difficulties posed by mass instruction at undergraduate level. Barker addresses weak points in the organization of Japanese university research, like the *kōza* system which makes it difficult to attain a critical mass of resources for major research projects.

However, there are also tendencies to strengthen universities' role as research institutions, as Yamamoto points out. Policy is moving toward a differentiation of universities into research universities and other types as a strategy to promote R&D by gradually transforming the present university structure. This has led to the appropriation of special funds to improve graduate schools, provide fellowships, and attract young scientists to university laboratories. It has to be seen to what extent these measures will prove suitable for improving the Japanese universities' role as research institutions which has been widely regarded as weak in the past.

The inter-organizational level

On the inter-organizational level, cooperative research among manufacturing firms is definitely increasing (Rokuhara 1985; Uesugi 1993). This reflects a global trend essentially caused by an increasing R&D cost burden for business firms. Many of them can no longer afford to conduct all their R&D activities only by themselves.

Inoue examines the case of Japan's semiconductor industry. He finds problems of increasing uncertainty and accessibility of semiconductor technology as the main motives of the recent drive in this industry to set up new R&D consortia. However, he also concedes that the R&D efforts of the US Semiconductor Manufacturing Technology Initiative (SEMATECH) as well as the increasing competition from countries just entering the semiconductor market – a situation Inoue describes as 'sandwiched' – has contributed to the invigoration of Japanese research efforts in the field of semiconductors. Inoue then extends his argument for cooperation aimed at countering the problems of uncertainty and accessibility in the development of semiconductor manufacturing technology to advocate increased international collaboration among nationally organized consortia.

In the public R&D sector, too, cooperation of researchers from different institutions has been facilitated. One reason for this is the gradual removal of administrative barriers for such cooperative projects throughout the last fifteen years, as analyzed by Barker; another reason is the setup of specific funds for cooperative research. While the organizational character of most of the R&D units is still bureaucratic, resulting in little incentive to cooperate with outsiders, collaborative projects among universities and public research institutions are gaining importance. Notably, Koji Omi, who was one of the key persons in the legislative initiative for the Science and Technology Basic Law, stressed the need to abolish regulative hurdles to collaborative R&D in a recent speech (Omi 1997).

Strengthening ties between industry and universities is a part of this policy which, in some areas, has led to a substantial flow of additional research funds into universities. Yamamoto concludes that the way toward multiple funding sources has thereby been opened for Japanese universities. However, he also points out that additional resources have to be complemented by a reduction in regulations, in order to allow the creative use of these resources.

The legally encouraged channel of public R&D funding for the private sector in Japan has been the establishment of a Mining and Manufacturing Technological Research Corporation (*Kōkōgyō Gijutsu Kenkyū Kumiai*) (Goto and Wakasugi 1988: 198–200). Whereas on the surface the legal form of such a venture stresses the cooperative aspect of publicly funded R&D, the administrative practice in the last three decades has mostly resulted in the exclusive support of R&D activities of a limited number of large corporations.

In Ray's paper the field of publicly funded research cooperation in Japan is discussed from the perspective of a comparison with the United Kingdom. It is argued that differences in this field have to be analyzed in the greater context of national innovation systems. In the case of Japan, strong government influence on certain economic activities and the system of life-time employment, which results in high inter-organizational barriers, appear to be of particular importance. Moreover, MITI-orchestrated research cooperation throughout the 1960s and 1970s has essentially been part of Japan's technological catch-up, whereas the governance structure of more recent projects has fundamentally changed, and their success has been much less spectacular compared with the earlier ones. The author concludes that in Western countries, lack of understanding of international differences in the institutional environment has resulted in a tendency to overestimate the synergy effects and to underestimate the governance problems of collaborative research.

While research cooperation has been gradually shifting away from those with a catch-up orientation to the front-runner types, this manner of public funding for business R&D activities is not questioned in itself. To the contrary, the increased government R&D budgets in the 1990s have resulted in the launch of several new research consortia. In the field of semiconductors, Inoue finds that these collaborative efforts differ from the famous Very Large Scale Integration (VLSI) consortium in several aspects. The present research consortia are largely initiated by industry. Moreover, MITI's own Association of Super-advanced Electronics Technologies (ASET) consortium, for example, is – because of the watchful eyes of foreign countries and World Trade Organization (WTO) rules – very much oriented toward basic research.

In other industries, the traditional tools of Japanese science policy (cooperative research among companies, funding of research and capital outlays, etc.) are also facing new challenges. In the case of Japan's pharmaceutical industry, such challenges are posed by the 'biological revolution.' Oberländer's paper finds that increased basic research has been recognized by the Ministry of Health and Welfare (MHW) as a key to innovation and as an essential requirement for the survival of the industry. However, in spite of numerous, although financially quite limited, research promotion programs of MHW, the shift toward more basic R&D has not happened. Although R&D investment in absolute terms and the number of researchers employed has almost doubled, research spending has not grown significantly relative to sales, and the dependency of Japan's pharmaceutical industry on foreign technology has not been reduced. The example of gene therapy, which is viewed by many as a key technology for future medical treatment, demonstrates Japan's slow reaction to new trends in the life sciences. It remains to be seen if the limited range of traditional Japanese science policy tools can be made suitable for the difficult R&D challenges of international networking and R&D risk management posed by the 'biological revolution.'

Nanotechnology is a field where no 'big bang' has happened either, as discussed by Sienko. This may be the reason why it is not yet even recognized as a field of research in its own right by most Japanese researchers active in it. The Japanese science bureaucracy, therefore, has no general plan for the future development of this field. Meanwhile, there exist numerous individual projects and programs sponsored by different government bodies which will give Japan a good chance to participate in cutting-edge research once it recognizes them as interconnected work. Especially in more application-oriented individual fields, Japanese industry appears to be well-positioned for a good start once a break-through occurs. However, the crucial link between 'dry' (based on the physical sciences) and 'wet' (based on the biological sciences) nanotechnology still appears to be missing in Japan. Besides, certain key technologies are not sufficiently developed – computer simulation for example. These difficulties seem to be related to the structural problems of Japanese science again: vertical organization, strict separation of organic and inorganic sciences, and turf-fighting among ministries.

Whereas Japan's response in some fields of high technology is rather slow, in others it may be described in contrast as going overboard. Becht addresses the field of advanced materials with special regard to superconductivity. Research in advanced materials as a whole has been subject to several programs of different Japanese government institutions since the 1980s. Superconductivity is a relatively new field of science, as it is essentially based on a scientific discovery that took place in 1986. Immediately after this discovery, several Japanese government institutions were eager to set up corresponding research units in an effort to take the international lead. A case study of the International Superconductivity Technology Center (ISTEC), one of these main units, reveals that a basic organizational feature is joint funding and dispatch of researchers by government and industry. While the project is not a complete failure, the fact that industry became involved in research funding when output was not yet commercially applicable turned out to be a problem. This has resulted in substantial frustration on the companies' side and a subsequent loss in their commitment. However, this may change in the future as the market for many of these materials is still rather limited, but is expected to expand rapidly in Japan as well as in the US.

A new channel for supporting private R&D with public money has recently been established in Japan: loan funds for venture businesses. In 1996, tens of billions of Yen have been budgeted by ministries, prefectures and local governments for this purpose. As it turned out, however, only a small fraction of these funds has actually been distributed to businesses (*Nihon Keizai Shinbun*, 18 January 1997). The main reason seems to be the absence of a 'venture culture' in Japan, resulting, among other factors, in a lack of entrepreneurs who have the ability to judge the innovative potential of small businesses. One possible way to remedy this is to attract large firms to invest in venture activities. To this end, the Japanese government is planning to relax anti-trust laws and deregulate the large corporations' ownership of venture stock (*Nihon Keizai Shinbun*, 17 February 1997).

As the public support of small companies' R&D itself is generally remarkable in Japan, persistent efforts are needed to create the *new* kind of innovative venture business culture. One way to do this is to promote the transfer of knowledge from the university laboratory to business. The Ministry for Education, Science and Culture (MESC) and the Ministry for International Trade and Industry (MITI) run such new programs: The Research for the Future Program is funded by MESC, and the Original Industrial Technology R&D Promotion Program is supported by MITI. The second program, which funnels MITI funds directly to university researchers, is especially noteworthy because it reaches across inter-ministry borders and may therefore be taken as a sign of new flexibility with regard to research funding (Normile 1996:1457–8). However, MESC is prepared to go even further. In the fall of 1997, the ministry was preparing a Bill to encourage universities to create 'technology liaison offices' fostering industry – university collaboration and to allow employees at national universities to participate in venture activities (Saegusa 1997: 105).

The international level

'Internationalization' has been a catch-phrase in Japan for a long time and the international involvement of the Japanese innovation system is rapidly progressing. Both Odagiri and Hemmert emphasize this issue as crucial for the transformation of Japan's industrial R&D. Manufacturing firms are increasingly hiring foreign specialists for their domestic R&D units and at the same time expanding their R&D activities abroad, namely in the US and Western Europe. Also remarkable is the sharp increase in 'strategic' foreign

R&D units that are not directly related to production or sales activities in the respective countries (Kiba 1996).

Barker combines intra- and inter-organizational aspects in his examination of the restructuring process in the Japanese science system with special regard to internationalization. His brief assessment of the current status of Japanese science shows that its degree of internationalization is much lower than in Western industrialized countries. This appears again to be related to fundamental weaknesses in its public research system, including low funding and a low level of reciprocation between research units and institutions. The author shows that since the 1980s various initiatives have been taken to improve conditions for public research in Japan, including programs to improve funding in certain promising areas, the removal of administrative barriers for inter-organizational reciprocation, and the fostering of an internationalization of science by foreign participation in domestic R&D and international collaboration programs. The paper reveals a strong interdependency between intra-organizational efforts for improving domestic research and inter-organizational ones aimed at increasing internationalization.

As a result, international activities are also rapidly gaining importance in the public R&D sector. While 'going abroad' has not been unusual for first-rate Japanese scientists and researchers in the past, Japanese research institutions have actively been recruiting foreign manpower only since the 1980s. This has resulted in a manifold increase of foreign researchers staying in Japan throughout the last decade, although their absolute number is still quite low compared to other industrialized countries (NISTEP 1995: 302–4). Other international activities include the launching and substantial funding of international R&D projects like the Human Frontier Science Program, and the setup and increase of fellowship programs for researchers from abroad, as described in Barker's paper.

Trends at the dawn of the twenty-first century: challenges and perspectives

In summary, with a view to their numerous predecessors, public programs to raise or at least maintain spending and advance a certain set of reforms aiming at stricter evaluation, result-oriented incentive structures and limited project duration in Japan's science and technology system appear to be of remarkable continuity. In the face of Japan's most recent economic downturn and with a view to R&D's role as a hoped-for source of economic growth, only the will or the ability to implement the above programs might have grown stronger. However, the struggle for fiscal reform and the campaign for administrative restructuring of the Japanese government add to the uncertainty about the continuity of this intention (Nathan and Triendl 1997: 815).

Concerning organization and management, internationalization has been by far the most dynamic development in the Japanese innovation system. Compared to earlier decades, there has also been some gradual change on the intra-organizational and inter-organizational levels, including innovations like new types of publicly financed research units and the setup of support programs for venture businesses. From the perspective of the Japanese innovation system, R&D operations abroad have been a novelty in many instances; however, they often provided good opportunities to compensate for weaknesses in the domestic system while avoiding reform of organizational structures and routines at home. Thus, the fundamental structure of most Japanese R&D units and the way their activities are governed seem to be mostly unchanged. The Japanese innovation system appears to follow a pragmatic pattern of development: old, existing organizational units and processes persist, while new ones with sometimes even radically different approaches are added, resulting in the coexistence of 'old' and 'new,' 'traditional' and 'innovative' organizational structures and processes. This makes it difficult to label the development as either fundamental change, gradual evolution or stagnation. To what degree old routines are allowed to persist and to what extent new structures are truly accepted will strongly influence how Japanese managers and scientists will master the perhaps greatest challenge that still lies ahead: to acquire the know-how necessary for dealing meaningfully with the globalization of research and development. Only by fully digesting the new ways in R&D management will it be possible to assess a potential domestic or international partner's 'technology seeds' and to form fruitful alliances in order to enhance one's research portfolio and manage development risk as will be essential for survival in the twenty-first century.

REFERENCES

Goto, Akira and Hiroyuki Odagiri (eds) (1997) *Innovation in Japan*. Oxford: Oxford University Press.

Goto, Akira and Ryuhei Wakasugi (1988) Technology Policy. In Komiya, Ryutaro, Masahiro Okuno, and Kotaro Suzumura (eds), *Industrial Policy of Japan*. Tokyo, San Diego and London: Academic Press, pp. 183–204.

Horres, Robert and Josef Kreiner (1992) Anspruch und Wirklichkeit der japanischen Forschungs- und Technologiepolitik – Eine Einführung. In *Trends der japanischen Forschungs- und Technologiepolitik. Miscellanea* 3. Tokyo: Deutsches Institut für Japanstudien, pp. 7–30.

Japan Economic Journal (4 December 1984), p. 1.

Kagaku Gijutsucho (1997) *Kuni no kenkyū kaihatsu zenpan ni kyōtsū suru hyōka no jisshi hōhō no arikata ni tsuite no taikōteki shishin* (National guideline on the method of evaluation for government R&D). Tokyo: Kagaku Gijutsucho.

Kagaku Gijutsu Kaigi (1996) *Kagaku Gijutsu Kihon Keikaku* (Science and Technology Basic Plan). Tokyo: Kagaku Gijutsu Kaigi.

Kiba, Takao (1996) *Nihon kigyō no kaigai ni okeru kenkyū kaihatsu no pafōmansu ni kan suru chōsa* (Survey of the performance of Japanese company R&D abroad). Tokyo: Kagaku Gijutsu Seisaku Kenkyusho (= NISTEP Report no. 43).

Nathan, Richard (1997) Japan's science spending climbs again. In *Nature*, vol. 385 (9 January), p. 104.

Nathan, Richard and Robert Triendl (1997) Radical restructuring in Japan merges science and education. In *Nature*, vol. 388 (28 August), p. 815.

Nihon Keizai Shinbun (30 December 1996), p. 13.

Nihon Keizai Shinbun (18 January 1997), p. 1.

Nihon Keizai Shinbun (17 February 1997), p. 3.

Nihon Keizai Shinbun (4 October 1997), p. 11.

Nikkei Weekly (10 June 1996), p. 3.

Nikkei Weekly (22 July 1996), p. 6.

NISTEP – Kagaku Gijutsu Seisaku Kenkyusho (1995) *Heisei 6-nenpan kagaku gijutsu shihyō. Nihon no kagaku gijutsu katsudō no taikeiteki bunseki* (Science and Technology Indicators 1994. A systematic analysis of science and technology activities in Japan). Tokyo: Kagaku Gijutsu Seisaku Kenkyusho (= NISTEP Report no. 37).

Normile, Dennis (1996) Japan hopes to cash in on industry–university ties. In *Science*, vol. 274, no. 5292. (29 November), pp. 1457–8.

NSF Tokyo Office (1996) Japan's Basic Law for Science and Technology. *Report Memorandum* no. 96–11. 3 April 1996.

Odagiri, Hiroyuki and Akira Goto (1993) The Japanese system of innovation: past, present, and future. In Nelson, Richard R. (ed.), *National Innovation Systems. A Comparative Analysis*. New York and Oxford: Oxford University Press, pp. 76–114.

OECD – Organization for Economic Co-operation and Development (1996) *Main Science and Technology Indicators 1995/2.* Paris: Organization for Economic Co-operation and Development.

Omi, Koji (1997) *The Present and Future Situation of Science and Technology Policy in Japan, Reforms Toward a Science and Technology-Oriented Nation.* Keynote Address, 7th International Forum on Technology Management, Kyoto, 4 November 1997.

Plate, Petra (1996) Der Grundlagenplan für Forschung und Technologie 1996: Eine Kurskorrektur. In *Japan Wirtschaft, Politik, Gesellschaft*, December 1996, pp. 659–65.

Rokuhara, Akira (1985) *Kenkyū kaihatsu to dokusen kinshi seisaku* (R&D and anti-monopoly policy). Tokyo: Gyosei.

Saegusa, Asako (1997) Japan ties the industry/university knot. In *Nature*, vol. 390 (13 November), p. 105.

Sakakibara, Kiyonori (1995) *Nihon kigyō no kenkyū kaihatsu manejimento* (R&D management in Japanese enterprises). Tokyo: Chikura Shobo.

Somucho Tokeikyoku (1997) *Heisei 7-nen kagaku gijutsu kenkyū chōsa hōkoku* (Report on the Survey of Research and Development 1995). Tokyo: Nihon Tokei Kyokai.

Tomita, Toshiki (1996) Facing up to Japan's impending fiscal crisis. In *NRI Quarterly*. Winter, pp. 2–17.

Triendl, Robert (1997a) Tighter evaluation urged for Japan's public research bodies. In *Nature*, vol. 387 (29 May), p. 444.

Triendl, Robert (1997b) Japanese budget austerity puts science plans at risk. In *Nature*, vol. 387 (12 June), p. 643.

Uesugi, Akinori (1993) *Tokkyō, nōhau raisensu keiyaku to kyōdō kenkyū kaihatsu* (Patent and know-how licensing contracts and collaborative R&D). Tokyo: Shoji Homu Kenkyukai.

Part II

INSTITUTIONAL STRUCTURE AND TECHNOLOGY POLICY

2

RESTRUCTURING BASIC, APPLIED AND DEVELOPMENTAL RESEARCH

Changes in allocation of R&D resources

Tomohiro Ijichi and Akira Goto[1]

Introduction

There seems to be a growing consensus in Japan that increased efforts in basic research are needed. This is reflected in the General Guideline for Science and Technology Policy, the document adopted by the Cabinet in 1986 and 1992, expressing an agreement by all ministries in the Japanese government on promotion of science and technology. The Science and Technology Basic Law that was passed in 1995 symbolizes a firm commitment toward the promotion of research and development (R&D) and determines its basic principles.

There are several reasons behind this enthusiasm for basic research, as will be discussed. They reflect the perceived change of Japan's status in a historical and international context. It is argued that the age of catching up is over and assumed that Japan is now at the forefront of the industrial and technological race. Thus, as the argument goes, creating rather than borrowing technology is more important, and for that reason, a basic research capability is the necessary prerequisite, especially in the high technology industries. It is also claimed that Japan, as one of the richest countries in the world, has an obligation to contribute to scientific knowledge as an 'international public good'.

The need to strengthen basic research capability has been emphasized in Japan for quite some time. In the 1980s, major R&D performing companies began to establish basic research laboratories, most of them independent of older, already existing laboratories. Although it is not clear whether R&D conducted at these laboratories was genuine basic research, the need for more long-range research with more ambitious targets focused on the early stage of innovation and the resulting demand for basic research was widely shared by the business community as well as by policy-makers. The Japanese government started various programs in 1981 to promote basic research, such as the Special Coordination Funds for Promoting Science and Technology, a budget[2] allocated for the purpose of international collaborative research, cooperation within industries, universities and government as well as vitalizing governmental institutes. The Exploratory Research for Advanced Technology (ERATO), a program[3] aimed at fostering the creation of advanced science and technology while stimulating future interdisciplinary scientific activities and improving conditions for basic research, also began in 1981. The Core Research for Evolutional Science and Technology (CREST), a competitive program[4] for research proposals referring to determined research areas in order to promote basic research in national institutes and universities by the Science and Technology Agency (STA), began in 1995. Again, even though some of the research projects under these programs may not necessarily be basic research in the strict sense of the term, a changing emphasis was apparent.

However, due to severe restrictions on government spending which finances a major part of basic research activity on the one hand and the rapid increase of R&D spending by the private sector in the 1980s on the other, the actual composition of R&D spending had not changed significantly.

Since 1992, as mentioned below, the expenditure for basic research has increased heavily as compared to applied research and development. This trend toward more basic research has been substantiated by the enactment of the Science and Technology Basic Law in 1995, which emphasized the need for more basic research as one of the most important goals of Japan's science and technology policy.

In this paper we will first examine the changing nature of R&D activity in Japan by using the widely used trichotomy of R&D: basic, applied, and developmental research. In the following section we investigate the composition of Japanese R&D and its change over time by using statistical data. We then review briefly the recent governmental science and technology (S&T) policies toward basic research in Japan, and finally discuss the background of changing composition of R&D in Japan and its implications for the future.

The changing nature of R&D in Japan

Japan's R&D expenditure reached its peak in 1992, and decreased thereafter for two consecutive years. This reflects the path of real gross domestic product (GDP) which grew only 0.1 percent in 1992 and 0.5 percent in 1993. It is the first time that R&D expenditure decreased since Japan began to collect national R&D data in 1960. Even when Japan experienced a negative growth rate of GDP in 1974 due to the oil crisis, R&D expenditure had increased. Until 1990, R&D expenditure grew faster than the GDP, resulting in an increase of R&D intensity. However, in the first half of the 1990s, the ratio of R&D expenditure to GDP, which reached 2.75 percent in 1990, gradually decreased.

Table 2.1 shows that the composition of R&D expenditures has changed in the 1990s. The ratio of R&D expenditure for basic research to total bottomed out in 1990 and has increased since. Notably, the absolute amount of Japan's spending on basic research is not small. In fact it is the second largest among the OECD countries, next to the United States and followed by Germany and France. As shown in Table 2.2, Japan's share of total expenditure for basic research in the OECD countries was 14.3 percent around 1990.

However, Japan's spending on basic research relative to its own total R&D spending has been and is still lower than that of other major R&D performing countries. Table 2.3 shows that the ratio of basic research to total R&D expenditure of Japan has been less than 15 percent, which is lower than that of the United States, Germany and France.

Who are the major performers of basic research in Japan? As Table 2.4 shows, while universities and colleges are the largest spenders on basic research, as in other countries, their relative share is significantly lower. While in other major R&D performing countries the share of basic research spending by universities and colleges to the total national basic research spending is more than 50 percent, the corresponding figure for Japan is only 46.8 percent. On the other hand, there is a high contribution by industry to national basic research in Japan (34 percent) compared to the United States (24.7 percent) and Germany (21.4 percent).

However, this is rather a recent phenomenon. The relative importance of sectors in performing basic research has changed over the last twenty years. As shown in Table 2.5, industry accounted for only about 19 percent of basic research spending in the second half of the 1970s. This ratio increased sharply and reached its peak of 40.6 percent in 1991, which was almost equal to the ratio of universities. Although decreasing since 1991, industry still accounts for about 34 percent in 1994. On the other hand, universities initially accounted for 64 percent in the second half of the 1970s, whereas their share decreased to less than 45 percent in 1991. In other words, the composition of basic research expenditure in Japan in the 1970s was almost the same as in other major R&D performing countries.

Why has companies' share in national basic research increased so much in Japan? As shown above, this development occurred in the early 1980s. Thus, the direct cause should be the hasty attempt to reduce the government deficit in the early 1980s. The reduction of government spending meant reduced spending for universities and colleges where a major part of R&D was basic research.

Aside from these historical reasons, there are several other possible explanations. As Rosenberg (1990) claims, firms conduct basic research with their own funds in order to monitor the forefront of science and engineering developed by universities and rival firms. Japanese firms, facing a competitive

Table 2.1 R&D expenditure by category of research in Japan (natural sciences only) (million Yen)

Fiscal year	*Total R&D expenditure*	*R&D expenditure specified by category of research*							
		Total		*Basic research*		*Applied research*		*Development*	
1976	2,941,373	2,895,139	100%	479,194	16.6%	713,801	24.7%	1,702,143	58.8%
1977	3,233,543	3,172,179	100%	513,914	16.2%	794,931	25.1%	1,863,333	58.7%
1978	3,569,953	3,507,996	100%	581,054	16.6%	879,908	25.1%	2,047,034	58.4%
1979	4,063,627	3,967,152	100%	620,728	15.6%	1,026,950	25.9%	2,319,475	58.5%
1980	4,683,768	4,538,392	100%	659,774	14.5%	1,153,393	25.4%	2,725,225	60.0%
1981	5,363,986	5,206,719	100%	724,348	13.9%	1,339,951	25.7%	3,142,421	60.4%
1982	5,881,539	5,795,048	100%	815,713	14.1%	1,498,871	25.9%	3,480,465	60.1%
1983	6,503,737	6,409,567	100%	896,693	14.0%	1,630,069	25.4%	3,882, 805	60.6%
1984	7,176,511	7,080,943	100%	959,941	13.6%	1,780,808	25.1%	4,340,194	61.3%
1985	8,116,399	8,018,275	100%	1,030,602	12.9%	2,001,771	25.0%	4,985,902	62.2%
1986	8,414,993	8,318,712	100%	1,102,843	13.3%	2,030,831	24.4%	5,185,038	62.3%
1987	9,016,186	8,914,221	100%	1,250,157	14.0%	2,166,770	24.3%	5,497,295	61.7%
1988	9,775,165	9,675,548	100%	1,287,965	13.3%	2,346,561	24.3%	6,041,022	62.4%
1989	10,909,335	10,827,318	100%	1,390,926	12.8%	2,588,408	23.9%	6,847,984	63.2%
1990	12,089,593	11,993,492	100%	1,510,178	12.6%	2,906,084	24.2%	7,577,231	63.2%
1991	12,720,140	12,617,443	100%	1,626,445	12.9%	3,111,792	24.7%	7,879,209	62.4%
1992	12,788,221	12,693,020	100%	1,711,645	13.5%	3,098,096	24.4%	7,883,279	62.1%
1993	12,528,354	12,418,908	100%	1,774,588	14.3%	2,990,615	24.1%	7,653,705	61.6%
1994	12,419,209	12,316,177	100%	1,781,854	14.5%	3,033,699	24.6%	7,500,623	60.9%
1995	13,191,200	13,083,400	100%	1,956,700	15.0%	3,217,700	24.6%	7,909,000	60.5%

Source: Somucho Tokeikyoku (1978–97).

Table 2.2 R&D expenditure among OECD countries (million Yen)

Country	*Year*	*Expenditure for basic research*		*Total R&D expenditure*	
Australia	1990	1,050.9	2.1%	3,662.7	1.2%
Austria	1989	350.8	0.7%	1,617.4	0.5%
Belgium	n.a.				
Canada	n.a.				
Denmark	1991	not specified		1,319.8	0.4%
Finland	n.a.				
France	1992	5,590.5	11.0%	26,584.0	8.5%
Germany	1991	6,628.6	13.0%	31,500.5	10.0%
Greece	1993	not specified		560.4	0.2%
Iceland	1992	16.1	0.0%	64.1	0.0%
Ireland	1990	26.7	0.1%	295.2	0.1%
Italy	1991	2,302.4	4.5%	12,986.7	4.1%
Japan	1989	7,301.3	14.3%	59,374.3	18.9%
Luxembourg	n.a.				
Mexico	n.a.				
Netherlands	1991	709.6	1.4%	4,740.2	1.5%
New Zealand	n.a.				
Norway	1993	232.8	0.5%	1,413.2	0.5%
Portugal	1990	103.5	0.2%	501.8	0.2%
Spain	1992	759.7	1.5%	3,816.7	1.2%
Sweden	1991	762.1	1.5%	3,806.4	1.2%
Switzerland	1992	277.0	0.5%	4,267.6	1.4%
Turkey	n.a.				
UK	n.a.				
US	1992	24,830.0	48.7%	157,400.0	50.1%
Total		50,941.8	100.0%	313,911.5	100.0%

Sources: OECD (1995a), (1995b).
Note: Purchasing power parities for GDP are used.

environment at home and abroad, may be keen to learn. Japanese firms might also have to conduct basic research by themselves because the research capability of Japanese universities is relatively weak. As mentioned above, Japanese universities lacked funds especially in the 1980s when the government tried to reduce its budget deficit. In addition, it is often criticized that its system is archaic and does not encourage innovative thinking. Recent research on spillovers from university research, for example Jaffe (1989), shows that firms benefit significantly from the research of local universities. Japanese firms may have felt handicapped in this regard and compelled to conduct basic research on their own.

To summarize:

1 Japan spent, and is still spending, less on basic research in relation to total R&D spending than other major R&D performing countries;
2 expenditure started to increase in the early 1990s;

Table 2.3 R&D expenditure by category of research in major industrialized countries (million Yen)

Country	*Total R&D expenditure*	*R&D expenditure specified by category of research*							
		Total		*Basic research*		*Applied research*		*Development*	
Japan (1994)[a]	12,419,209	12,316,177	100%	1,781,854	14.5%	3,033,699	24.6%	7,500,623	60.9%
United States (1994)[c]	30,951,000	30,951,000	100%	5,350,360	17.3%	7,194,750	23.2%	18,405,890	59.5%
Germany (1991)[b,c]	6,810,881	6,046,201	100%	1,272,288	21.0%	4,773,913	79.0%		
France (1992)[c]	4,947,335	4,947,335	100%	1,040,395	21.0%	1,557,918	31.5%	2,349,022	47.5%

Sources: National Science Foundation (1995); OECD (1995b); Somucho Tokeikyoku (1996).

Notes:

a R&D expenditure in Japan is only for natural sciences.

b In Germany, applied research and experimental development are not distinguished.

c Purchasing power parities for GDP are used.

Table 2.4 Expenditure for basic research by performing sector in major industrialized countries (million Yen)

Country	*Total*		*Industry*		*Private research institutions*		*Governmental research institutions*		*Universities and colleges*	
Japan (1994)[a]	1,781,854	100%	606,380	34.0%	121,845	6.8%	218,992	12.3%	834,637	46.8%
United States (1994)[b]	5,279,770	100%	1,303,200	24.7%	392,770	7.4%	488,700	9.3%	3,095,100	58.6%
Germany (1991)[b]	1,272,288	100%	271,641	21.4%	4,296	0.3%	309,755	24.3%	714,017	56.1%
France (1992)[b]	1,040,395	100%	136,907	13.2%	26,234	2.5%	194,981	18.7%	682,274	65.6%

Sources: National Science Foundation (1995); OECD (1995a), (1995b); Somucho Tokeikyoku (1996).

Notes:

a R&D expenditure in Japan is only for natural sciences.

b Purchasing power parities for GDP are used.

Table 2.5 Expenditure for basic research by performing sector in Japan (million Yen)

Fiscal year	*Total*		*Industry*		*Private research institutions*		*Research institutions*		*Governmental research institutions*		*Universities and colleges*	
1976	479,194	100%	94,858	19.8%	..		77,917	16.3%	..		306,420	63.9%
1977	513,914	100%	98,761	19.2%	..		84,692	16.5%	..		330,461	64.3%
1978	581,054	100%	106,486	18.3%	..		98,847	17.0%	..		375,721	64.7%
1979	620,728	100%	122,886	19.8%	..		110,086	17.7%	..		387,756	62.5%
1980	659,774	100%	157,413	23.9%	..		103,852	15.7%	..		398,509	60.4%
1981	724,348	100%	189,297	26.1%	..		105,880	14.6%	..		429,171	59.2%
1982	815,713	100%	221,386	27.1%	..		110,952	13.6%	..		483,375	59.3%
1983	896,693	100%	259,849	29.0%	..		112,423	12.5%	..		524,421	58.5%
1984	959,941	100%	290,030	30.2%	..		124,464	13.0%	..		545,447	56.8%
1985	1,030,602	100%	351,657	34.1%	..		131,503	12.8%	..		542,442	52.6%
1986	1,102,843	100%	371,257	33.7%	..		156,903	14.2%	..		574,683	52.1%
1987	1,250,157	100%	429,203	34.3%	..		201,165	16.1%	..		619,789	49.6%
1988	1,287,965	100%	474,913	36.9%	73.450	5.7%	..		116,916	9.1%	622,686	48.3%
1989	1,390,926	100%	524,321	37.7%	86,343	6.2%	..		120,206	8.6%	660,056	47.5%
1990	1,510,178	100%	589,072	39.0%	88,139	5.8%	..		131,760	8.7%	701,207	46.4%
1991	1,626,445	100%	660,219	40.6%	95,814	5.9%	..		145,112	8.9%	725,300	44.6%
1992	1,711,645	100%	656,014	38.3%	93,952	5.5%	..		184,395	10.8%	777,284	45.4%
1993	1,774,588	100%	605,047	34.1%	111,927	6.3%	..		230,279	13.0%	827,335	46.6%
1994	1,781,854	100%	606,380	34.0%	121,845	6.8%	..		218,992	12.3%	834,637	46.8%

Sources: Somucho Tokeikyoku (1978–96).

3 the absolute amount of Japan's spending on basic research is the second largest among OECD countries;
4 although the government is the largest funding source, the private sector spends more than that in other major R&D performing countries relative to total R&D; and
5 the above-mentioned facts are related to the smaller role of the government with respect to funding of R&D compared to other major R&D countries.

Recent government policies to promote basic research

In this section, recent government policies toward basic research are briefly reviewed. First, the Japanese government has emphasized the need to promote basic research since the middle of the 1980s. The General Guideline for Science and Technology Policy, which expresses an agreement of all ministries in the Japanese government to promote science and technology, was adopted by the Cabinet in 1986 and 1992. The General Guidelines had been prepared by the Council for Science and Technology (CST), the advisory and inter-ministerial coordinating body on S&T policy at the highest level. One of the recommendations submitted by CST on request of the Prime Minister advised that the focus of research at national research institutes should shift from testing and development research to basic research.

Second, in order to channel more funds for basic research into national research institutes, the Japanese government has formulated various kinds of national programs within the framework of vitalizing national research institutes in the Special Coordination Funds for Promoting Science and Technology. They include:

- the Centers of Excellence (COE) nurturing program (started in fiscal year 1993, 3,600 million Yen in fiscal year 1996 budget), a program of increasing funds for the national institutes which aim at COE;
- the inter-ministry basic research program (started in fiscal year 1988, 800 million Yen in fiscal year 1996 budget), a program to promote cooperation among national institutes on basic research beyond boundaries between ministries; and
- the basic research program on prioritized areas (started in fiscal year 1985, 2,100 million Yen in fiscal year 1996 budget), a program of promoting basic research to create technological seeds in national institutes.

Third, in order to increase domestic demand and reduce the trade surplus, the Japanese government promised to increase public spending, and as a result, the amount of R&D spending has increased significantly in supplementary budgets since fiscal year 1992.

Fourth, the Japanese government began to classify R&D expenditure by government as public investment rather than current expenditure. This meant that it became possible to finance public R&D through government bonds. This new treatment of government spending on R&D opened the way for increasing national R&D investments and enabled the national government to finance special corporations (*tokushu hōjin*) or authorized corporations (*ninka hōjin*) which control a large amount of governmental R&D investment. For example, the Science and Technology Agency (STA) formulated the CREST program (started in fiscal year 1995, 12,846 million Yen in fiscal year 1996 budget) by using the Research and Development Corporation of Japan (JRDC), one of the special corporations affiliated with the STA. Other ministries such as the Ministry of International Trade and Industry (MITI), the Ministry of Education, Science and Culture (MESC) and the Ministry of Health and Welfare (MHW) also started similar programs using the similar schemes as the STA.

Finally, the Cabinet adopted the Science and Technology Basic Plan in July 1996 which replaced the former General Guideline for Science and Technology Policy. The Science and Technology Basic Law, passed by the Diet in November 1995, stipulates that the government shall establish the Science and Technology Basic Plan for executing general measures to promote science and technology. It resembles a five-year plan with prospects to the next decade. The Science and Technology Basic Plan of 1996 stated that the total amount of government-funded R&D expenditures should be about 17 trillion Yen for a five-year period between fiscal year 1996 and fiscal year 2000. This means that the Japanese government should double government-funded R&D expenditure by fiscal year 2000 as compared to fiscal year 1992. Moreover, the Science and Technology Basic Plan of 1996 states that the government should promote basic research more intensively.

Discussion

In this section, we discuss the reasons usually given when the need to strengthen basic research in Japan is emphasized, and, moreover, their implications for the future with respect to the current R&D situation.

First, it is claimed that Japan, one of the richest countries in the world, should contribute more to scientific research which increases the pool of human knowledge as an 'international public good'. Japan's expenditure for basic research relative to its total R&D is somewhat lower than that of other R&D performing countries as we have seen. However, as shown in Table 2.2, Japan's contribution to world basic research in *absolute* terms of expenditure is rather large. Concerning the output of basic research, Japan's share of publications in scientific papers is second only to the United States, according to the Science Citation Index Database. In addition, Japan is contributing to international mega-science projects such as CERN, and has initiated large-scale international projects such as the Human Frontier Science Program. This indicates that Japan is already a major contributor to the world scientific community.

As mentioned earlier, it seems that government spending on basic research will increase. To supplement these spending efforts, various aspects of R&D management, such as organizational restructuring and allocation of funds to higher education and national institutes, have to be taken into consideration. It is often reported that the productivity of national institutes and universities is constrained under the current legislation on governmental organizations, for instance by the rigid budgeting system that allows to plan only for single years, the detailed, predetermined and unchangeable items of expenditure, the restriction on numbers of researchers and research assistants, and the weak responsibility and authority of management. Some observers believe that national institutes and national universities should be privatized or changed into special corporations such as the Institute of Physical and Chemical Research (RIKEN), one of the most widely recognized institutes in Japan.

Second, it is argued that as Japan moved toward the technological frontiers, technological 'seed' is becoming scarce. This is based on the notion that Japan has only been an importer of technological seeds in the past, which were then improved and commercialized. This is becoming more and more difficult, as the argument goes, partly because available 'seeds' from abroad are exhausted as Japan reaches the frontiers, and partly because foreign countries have become reluctant to give 'seeds' to Japan.

It is difficult to determine whether technological 'seeds' have indeed become scarce. This type of concern has been voiced for a long time. For instance, the Economic Survey of Japan, the economic White Paper of 1956 declared that the postwar period was over and that Japan had to develop an 'indigenous technology' as the technological seeds were becoming scarce and difficult to obtain. Japan, which had been importing technology from abroad in the prewar decades, became isolated during the Second World War. Thus, Japan actively imported this backlog of technology in the 1950s and early 1960s. However, technology imports were increasing since then, as indicated by statistics on technology trade. On the other hand, as the patent data show, Japan itself became a major inventor. Some economists, notably Evenson (1993), argue that technological opportunity is decreasing world-wide and that this is the cause of decreasing patent productivity (number of inventions per R&D expenditure) throughout major industrialized countries except Japan. Several alternative hypotheses were presented and the real cause of declining patent productivity is not yet really known. But we do know that patent productivity has not declined in Japan.

It is often argued that Japan now stands at the technological frontier. However, according to the Delphi Survey (NISTEP and FhG-ISI 1994), as shown in Figure 2.1, the technological level in many fields is assessed by Japanese scientists and engineers as inferior to that of other countries. Thus,

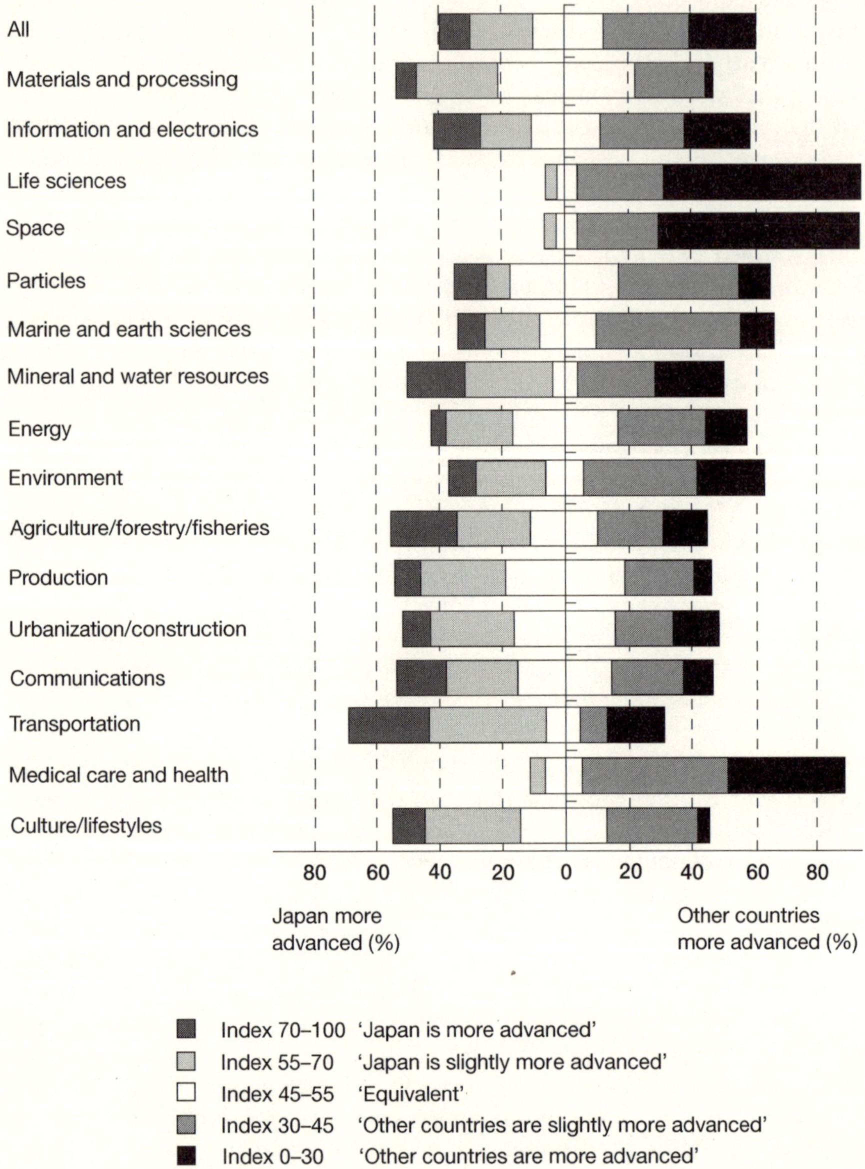

Figure 2.1 Trends in each technological area comparing the current R&D level of Japan and other countries

Source: NISTEP and FhG-ISI (1994, figure 2.6.2).

it remains an important issue that advanced technologies as well as basic research should be actively promoted.

Third, it is often claimed that high-technology industries may generate externalities which spill over within national boundaries far more easily than across borders. Thus national welfare could be improved by the promotion of high technology industries. If this argument holds, high technology industries should be promoted, and, as high technology industries are science-based, it is necessary to promote science or basic research.

High technology industries and science are linked in many ways. First, in high technology industries, scientific discoveries often lead to direct innovation. The classic example is Cohen-Boyer's discovery of recombinant DNA in 1973 and the following emergence of biotechnology industries. This reasoning is based on a linear model which is often criticized as too simplistic, but the linear model may be relevant in high technology industries because the linkage between science and innovation is more direct than in other industries. Second, as Rosenberg and Nelson (1994) and others emphasize, basic research provides a pool of knowledge which is useful for problem solving in the industry. This is particularly relevant to high technology where problems to be solved require the understanding of sophisticated leading edge science.

Third, the understanding of leading edge science is also necessary to monitor R&D by universities and rival firms in order to compete in the high technology industries. As mentioned earlier, evidence is given in recent empirical studies such as Jaffe (1989), which showed that the existence of a university resulted in more patents from firms or start-up of new companies in the same regional area.

Despite the above argument, Japanese industry has decreased expenditures for basic research since 1991. This decline may result in a decrease of advantages in high technologies in the future. Thus, basic research undertaken by industry may be promoted as well as by universities and research institutes.

NOTES

1 The views expressed in this chapter are solely those of the authors and do not represent the view of the National Institute of Science and Technology Policy, Science and Technology Agency.

2 This budget is not allocated through each ministry such as usual governmental budgets but through the Science and Technology Agency (STA) based on the selection by the Council for Science and Technology (CST), which is assigned for coordination among ministries.

3 In the ERATO program, project leaders who are selected by the Japan Science and Technology Corporation (JST), one of the special corporations affiliated with STA, set up research themes, select about fifteen researchers, and conduct the projects for five years with a budget of about 2 billion Yen per project.

4 In the CREST program, JRDC accepts applications of research proposals for

determined research areas from researchers, and selects the research leaders and research themes from the applications.

REFERENCES

Evenson, R.E. (1993) Patents, research and development, and invention potential – international evidence. In *American Economic Review*, vol. 83, pp. 463–8.

Jaffe, A.B. (1989) Real effects of academic research. In *American Economic Review*, vol. 79, pp. 957–70.

National Science Foundation (1995) *National Patterns of R&D Resources*. Washington, DC: National Science Foundation.

NISTEP and FhG-ISI – National Institute of Science and Technology Policy and Fraunhofer Institute for Systems and Innovation Research (1994) *Outlook for Japanese and German future technology: Comparing Japanese and German technology forecast surveys.* Tokyo: National Institute of Science and Technology Policy and Fraunhofer Institute for Systems and Innovation Research (= NISTEP Report no. 33).

OECD – Organisation for Economic Co-operation and Development (1995a) *Basic Science and Technology Statistics*. Paris: Organisation for Economic Co-operation and Development.

—— (1995b) *National Accounts 1960–1993*. Paris: Organisation for Economic Co-operation and Development.

Rosenberg, N. (1990) Why do firms do basic research with their own money? In *Research Policy*, vol. 19, pp. 165–74.

Rosenberg, N. and R.R. Nelson (1994) American universities and technical advance in industry. In *Research Policy*, vol. 23, pp. 323–48.

Somucho Tokeikyoku (1978–1997) *Kagaku gijutsu kenkyū chōsa hōkoku* (Report on the Survey of Research and Development). Tokyo: Nihon Tokei Kyokai.

3

THE INTERACTION BETWEEN TECHNOLOGY AND ECONOMY

Has the 'virtuous cycle' of Japan's technological innovation system collapsed?

Chihiro Watanabe and Martin Hemmert

Introduction

The remarkable development of the Japanese economy throughout the last fifty years has largely been attributed to the driving force of industrial development and constant efforts to increase technological innovation. To date, a number of authors have identified the sources supporting Japanese industry's technological progress (Mowery and Rosenberg 1989; US Department of Commerce 1990; Odagiri and Goto 1993). None, however, have scrutinized the aspect of positive feedback loops between technological innovation and economic development. In the postwar decades, an elaborate array of several reinforcing factors have evolved in Japan, leading to a 'virtuous cycle' (i.e. successful stimulation and inducing interaction) between technology and economic development (Watanabe 1995a). This mechanism, as a result of combined industry efforts and government stimulation, functioned quite well during the serious energy and environmental crises of the 1970s and early 1980s (Watanabe and Honda 1992). The MITI (Ministry of International Trade and Industry) that was mainly responsible for industrial technology policy, stimulated and induced industry's efforts in particular by establishing a policy system which has strengthened the dynamism of technological development (Watanabe, Santoso and Widayanti 1991; Watanabe and Clark 1991).

However, since the relaxation of energy constraints (starting in 1983), the sharp appreciation of the Yen (triggered by the Plaza accord in 1985), the succeeding 'bubble economy' (1987–90) and its bursting (1991), Japanese industry has faced a structural stagnation of R&D activities which may result in the collapse of the virtuous cycle between technology and economic development (Watanabe 1995b). Once such a cycle has collapsed, its reactivation becomes difficult. Thus far, Japan has paid little attention to this aspect, leading to insufficient empirical analyses of the impacts of stagnation of R&D activities on Japanese industry.

Focusing on the perspective of the positive interaction between technological innovation and economic development, this paper analyzes the source of this virtuous cycle in the Japanese manufacturing industry as well as the role of the policy system and its mechanism. In addition, it reviews the current concerns regarding the cycle's collapse. In methodological terms, the findings are based on an analysis of the relationship between aggregated indicators of technological and economic activity. This aggregated approach has a number of shortcomings such as measurement problems and the possibility of a lack of validity of statistical relationships that are assumed. However, it follows a main branch of aggregated economic analysis, i.e. growth accounting (Jorgenson 1995) and also offers the chance to gain insights into the overall structure of the technological and economic system and its evolution that cannot be drawn from field-specific work.

This paper addresses two basic hypotheses. First, the dynamism of postwar economic development in Japan was sustained through a strong positive interaction between capital investment and technological advance. Second, this interaction has weakened since the period of the 'bubble economy,' when R&D activity in Japanese industry began to stagnate. It is organized as follows: the next section gives an empirical review of Japan's path of economic development and technology's contribution to this development. For analytical purposes, this mechanism is broken down into the complementary relationship between technology and capital and the substitution of constrained production factors by technology. The third section attempts an empirical analysis of the impact of institutional surroundings such as the economic, social and cultural environment and of science and technology policy on this development and its mechanism. In the fourth section, the current R&D stagnation in Japan and its background is analyzed while the final section reviews the findings concerning Japan's technological innovation system and draws some further implications for the future direction of R&D.

The role of technology in Japan's economic development

The complementary relationship between technology and capital

The Japanese economy has shown tremendous growth during the last decades due to its industrial development. Japan's world GDP share was 4.1 percent in 1960, 6.4 percent in 1970, and 9.1 percent in 1980. It increased to 14.8 percent in 1990 and was more than 18 percent in 1995[1] (United Nations 1996). This outstanding growth performance can be attributed largely to rapid technological progress as demonstrated in Figure 3.1.

Whereas agriculture, forestry, fisheries and mining generally stagnated in the postwar period, manufacturing industry took a leading role in stimulating Japan's economy as a whole (Keizai Kikakucho 1991; Watanabe 1990). Its share of national employment was rapidly increased until the 1980s. Moreover, the manufacturing sector, and within this sector particularly the machinery industry, generated persistently high growth rates in value added products, and therefore accounted for an increasingly large share of Japan's GDP. The outstanding aggregated performance of the manufacturing industry was a result of the high dynamism and initiative of manufacturing companies in shedding obsolete equipment, facilities and technology, and venturing into new lines of activity, all of which rapidly enhanced technology and productivity levels (Tsusho Sangyosho 1988). These efforts resulted in the attainment of levels outperforming foreign competitors and recognition as being among the world's most advanced nations (US Department of Commerce 1990).

Such remarkable improvement has mainly resulted from private industry's vigorous efforts to invest into R&D. The R&D intensity (relation between R&D expenditures and sales) in the Japanese manufacturing industry rose from less than 1.5 percent in the 1950s and 1960s to around 3.5 percent in the early 1990s (Somucho Tokeikyoku 1996).

In order to clarify the relationship between the rapid growth of the manufacturing industry as a whole and the fast expansion of R&D efforts, two observations are crucial. First, the marginal productivity of Japanese industry's capital investment has exceeded those levels found in the US and European countries (Yoshikai 1985). Second, the marginal productivity of its R&D investment (rate of return to R&D investment) has proven to be much higher than that of capital investment, resulting in an internal rate of return on R&D investment much higher than in other advanced countries (Nihon Kaihatsu Ginko 1984; Goto 1993: 44–8). One possible explanation is that as Japan was in a technological catch-up process to the West, high productivity gains could be achieved with relatively limited R&D expenditures.

This high rate of return to R&D investment in Japan's industry induced further efforts by private industry to increase R&D investment. It is important to note that these efforts were incorporated with capital investment (Yoshikai 1985; Watanabe 1995a), resulting in a virtuous circle between economic growth (incorporated in capital investment) and technological dynamism (incorporated in R&D expenditures).

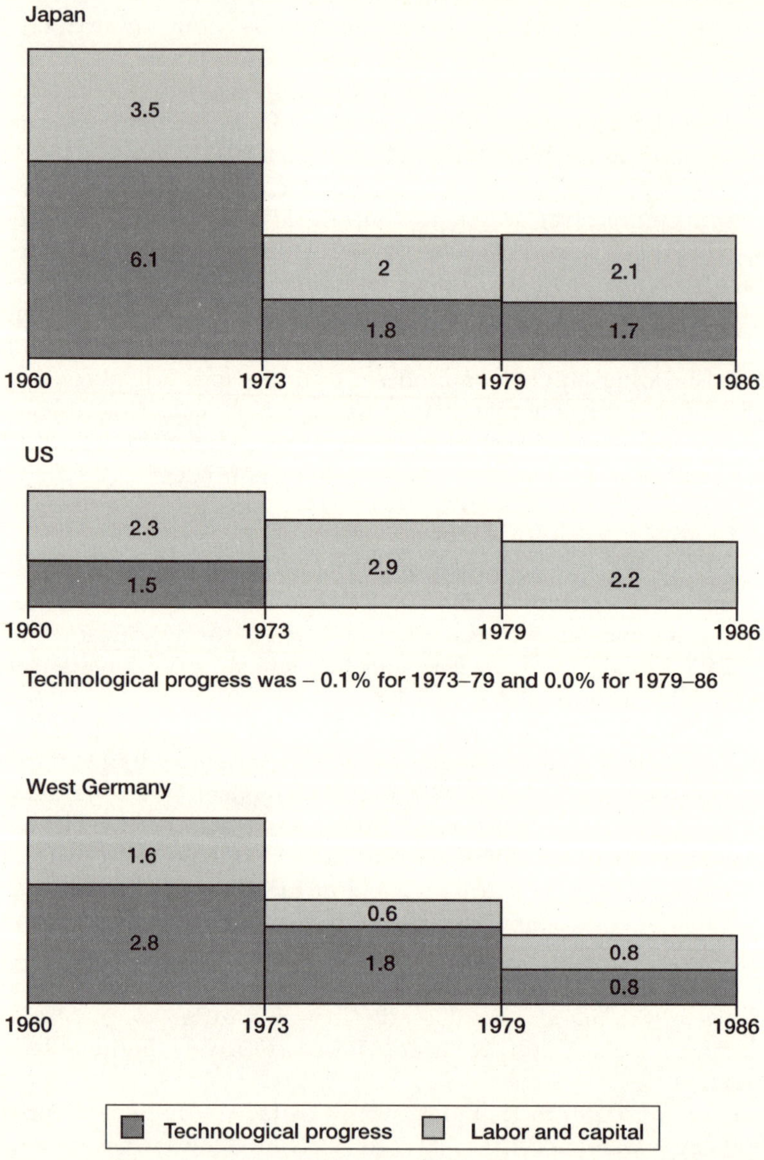

Figure 3.1 Comparison of the contribution of technological progress to economic growth in Japan, the US and West Germany 1960–86 (annual rates, percent)

Source: Englander and Mittelstädt (1988).

Note: The contribution of technological progress is measured by Total Factor Productivity (TFP).

Technology substitution for constrained production factors

Despite many handicaps, Japan achieved a rapid enhancement of its technology and productivity levels by focusing its efforts on improving the productivity of relatively scarce resources (constrained production factors) in every era. Although capital was a scarce resource up until the 1950s, with economic development, the bottleneck shifted to labor in the 1960s, environmental capacity from the mid-1960s to the start of the 1970s, and energy following the first energy crisis in 1973. While this development has often been explained by the complementary relationship between R&D and capital investment, technology in fact provided the strongest contribution through its substitution for scarce resources. Noteworthy is that the marginal productivity of technology fell below the productivity of capital from 1973–83. Nevertheless, industry's efforts in consistent R&D investment were sustained also during this period. This is considered to be due to industry's efforts to substitute technology for energy, a crucially constrained production factor during the period between 1973 and 1983.

Figure 3.2 illustrates trends in complementary and substitutional relationships between labor, capital, energy and technology (technology knowledge stock) in Japan's manufacturing industry from 1956 to 1992. We note that technology and capital were consistently complementary until the late 1980s; technology consistently substitutes for labor; while energy and technology were independent or slightly complementary until 1973 when technology began to substitute for energy. These trends confirm the hypothesis

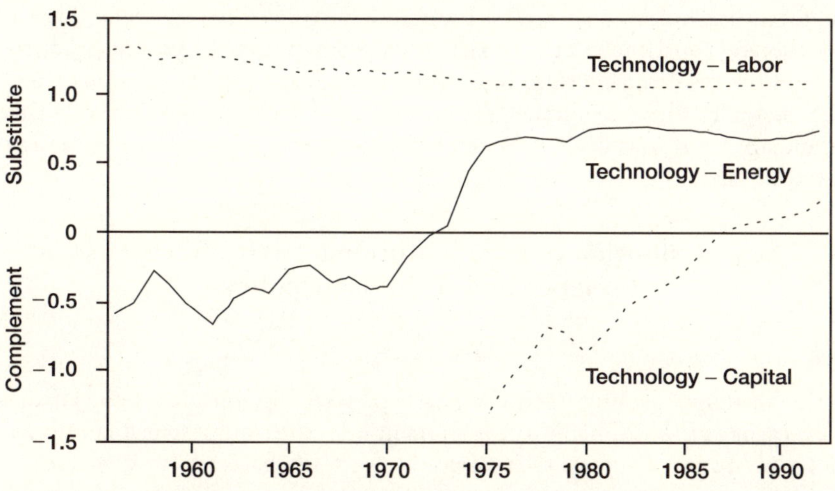

Figure 3.2 Trends in substitution and complement between labor, capital, energy, and technology in the Japanese manufacturing industry, 1956–92 (Allen partial elasticity of substitution)

Source: Authors' calculations.

that there was a strong positive interaction between capital investment and technological progress.

Facing the energy crises in the 1970s, with the support of the above mechanisms, Japan was able to largely reduce the energy intensity of its economy. Whereas the value added in Japan's manufacturing industry more than doubled between 1974 and 1994, energy consumption and CO_2 emission have been reduced during the same period.

An econometric analysis of the factors influencing the trend of CO_2 emissions between 1974 and 1994 reveals that improvement in energy efficiency contributed 70 percent to the overall reduction of CO_2 emission. In fact, the performance of Japan's manufacturing industry in improving energy efficiency was outstanding in comparison to other advanced countries (OECD 1991).

This remarkable improvement in energy efficiency was mainly the result of mutually stimulating energy R&D and non-energy R&D efforts. It was directly achieved by the development of energy substituting technologies such as advanced energy conservation technologies and energy efficient production technologies, and indirectly by introducing advanced energy conservation facilities and less energy dependent production processes. This indicates that technology made a significant contribution to improving the productivity of scarce resources, constrained energy and environmental capacity. Japan's manufacturing industry was thereby able to overcome energy and environmental constraints while maintaining sustainable development.

Thus, despite numerous handicaps, through the support of the complementary relationship between technology and capital as well as technology substitution for scarce resources (constrained production factors), Japan's manufacturing industry showed a high ability to improve its technological level, thereby contributing to the rise in its productivity. Improved productivity and the resulting increase in production induced further vigorous R&D which again resulted in further enhancement of technology. Through this mechanism, a virtuous cycle between technology and economic development evolved in Japan.

The contribution of the institutional environment and of science and technology policy

Overview

Over the last four decades, technology has played a significant role in Japan's achievement of sustainable development despite numerous handicaps. Among production factors, technology has certain characteristics, including intangibility, uncertainty, high risk, high cost, and a long lead-time. These characteristics have two important implications. First, private industry generally flinches from challenging technological investment without certain favorable conditions. Second, technology can maximize its potential performance only in a socio-economic system that fundamentally supports its development.

As will be argued subsequently, Japan's success in constructing a virtuous cycle between technology and economic development can be attributed to a sophisticated combination of industry efforts and government stimulation. Government stimulation was focused on constructing a socio-economic system in which technology could maximize its potential performance. Such a system is dynamic, comprehensive and organic, corresponding to both domestic and international environments in each respective era.

The economic environment and social and cultural foundations

In 1988, at a time when the competitive strength of Japan's manufacturing industry seemed to be overwhelming, the results of two surveys were published that analyzed the trend of Japan's technological level compared with the US in the five-year period between 1982 and 1987 concerning forty-one leading high-technology products (Tsusho Sangyosho 1988). The report revealed that Japan's relative position greatly improved during that period in many fields that were surveyed, and identified the following factors as important in contributing to the rapid improvement of Japan's high-technology products:

- *Severe, productively directed competition.* Domestic and international competition in several markets, such as amorphous alloys, fine ceramics, and semiconductor lasers have been very vigorous. However, cut-throat competition was avoided, resulting in a productive, rather than destructive overall effect on industry performance.
- *High quality needs of users and consumers.* In a number of industries, including polymer separation membranes, advanced composite materials, and digital private branch exchanges (D-PBX), high quality requirements by end-users and consumers, had a significant stimulating effect.
- *Inter-industry stimulation.* The development and application of related technologies led in industries like spectrum analyzers, laser printers and charge coupled devices (CCD) to a positive impact on technology development, also by stimulating competition.
- *Mutual stimulation between changes in industrial structure and advancements in R&D.* Most typically observed in this category are particle accelerators, communication satellites and magnetic resonance imaging (MRI).
- *Progress in basic and original research thinking.* The development of products such as microprocessors, bio-products, and super high-rise structures has been strongly supported by this.
- *Qualifications and attitudes of workers and high standards of quality control.* A significant positive impact of these factors was observed in semiconductor memory devices and computers.

- *Stimulation provided by national R&D projects.* The development of products such as laser processing machines, aircraft engines, and photo-voltaic power generation equipment was significantly fostered by such projects.

It goes without saying that the results of this survey are not necessarily representative for the determining factors of the whole of Japanese industrial R&D in the last five decades, as it covered only the development in a limited number of industries throughout the 1980s. However, it reveals a number of hints by what environmental conditions R&D in Japanese manufacturing was fundamentally supported. These factors can be divided into economic, socio-cultural, and international ones.

Concerning economic conditions, the high competitive pressure Japanese manufacturing companies have steadily been confronted with throughout the past decades must be regarded as a key factor when explaining their vigorous R&D efforts (Odagiri 1994). One supporting fundamental here is the large size of the Japanese market that forestalled natural monopolies in most markets for high-tech products. Another is the method of *how* companies have been exercising competition. Compared with price competition, quality competition appeared to play a relatively large role among Japanese manufacturers, resulting in a high stimulus to develop new products and apply new production processes.

Moreover, the institutional setting for technological diversification of industrial companies further stimulated R&D activities. As the conditions for mergers and acquisitions have been highly unfavorable in Japan, firms diligently tried to expand into new markets by applying their own technological knowledge to new products (Goto 1993: 61–7). This resulted in frequent R&D driven market entries.

Another crucial environmental factor is the educational level of the working population. A high level of education is a fundamental requirement for a society which demands high quality goods. The general level of education, although not that of professional skill formation, has traditionally been very high in Japan, providing a good setting for R&D activities. The high commitment of workers and managers is another supporting factor.

The economic environment and social and cultural foundations coincided with a favorable international environment which contributed to Japan's economic development after the Second World War. The progressing liberalization of international trade under the GATT regime allowed Japanese manufacturers to expand into foreign markets, stimulating further capital investment and technological progress. In the 1970s and 1980s, a number of external shocks, namely the energy crisis and Yen crisis (the appreciation of the Yen), has shaken the Japanese economy. The efforts to overcome these crises, however, have resulted in a further strengthening of the international competitiveness of Japanese industry via the substitution of constrained factors by technology, as shown above.

In the preceding paragraphs, a systematic view of the factors facilitating dynamic industrial development in Japan has been developed. In this context we note that on the basis of favorable economic, social and cultural conditions, a dynamic development of industry and technology could take place. Under such conditions, the role of government policy was to foster the industrial dynamics. In Japan, this could be achieved with financially limited measures, as will be shown in the next section.

The role of science and technology policy

Figure 3.3 illustrates trends in Japan's governmental support for R&D investment by industry. We note that Japan's governmental R&D funding represented 5–10 percent of total industry R&D expenditures by the mid-1960s. As Japan's economy expanded, the relative level of government R&D funding decreased and currently it is only approximately 3 percent. Interestingly enough, Japan's governmental support for R&D investment by industry is extremely small compared to that of other advanced countries, as summarized in Table 3.1. The ratio of government R&D funding as a portion of all national R&D expenditures is also much smaller in Japan than in other advanced countries.

In order to attain a proper understanding of the system of Japanese science and technology policy, one first has to notice that competence and budgets in

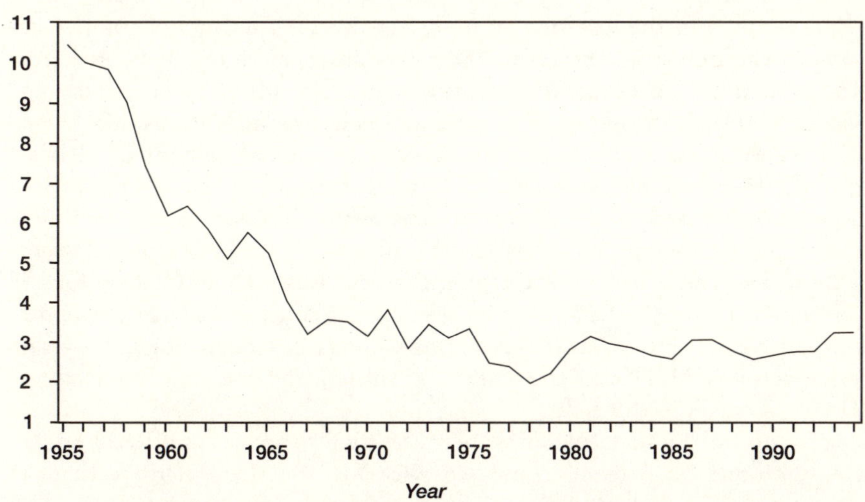

Figure 3.3 Japan's governmental support for industrial R&D development, 1955–94 (government funding as a portion of the total industrial R&D expenditure, percent)

Sources: Wakasugi (1986); Kagaku Gijutsucho (Science and Technology Agency) (1996 and various years).

Table 3.1 Government R&D spending in advanced countries

Country	*Japan (1994)*	*US (1995)*	*Germany (1995)*	*France (1992)*	*UK (1993)*
Proportion of government funding of industry's R&D expenditures (%)	3.0	20.2	9.0	8.6	13.5
Proportion of government funding of national R&D expenditures (%)	20.2	35.5	39.3	44.3	32.3

Sources: Kagaku Gijutsucho (Science and Technology Agency) (1996); BMBF (1996).
Note: Government funding of industry's R&D expenditures includes funding of private research institutions.

this field are shared by several institutions. The three most important actors are the Ministry of Education, Science and Culture (MESC), the Science and Technology Agency (STA), and the Ministry of International Trade and Industry (MITI). Roughly speaking, the following division of labor has been established between the three ministries. MESC has the administrative competence for the academic research system, namely national universities. STA focuses its activities on large-scale research projects in government-dominated sectors like atomic energy and space development. Finally, MITI has its domain in the fostering of industrial R&D. The lines of competence between the ministries, however, are not as clear-cut as might be expected. Moreover, other government institutions like the Ministry of Health and Welfare (MHW), the Ministry of Posts and Telecommunications (MPT), and the Defense Agency also play a major role in fields of technology that are strongly related to their core activities. Altogether, it can be fairly stated that in Japan there exists no single government institution that widely dominates science and technology policy as a whole such as the Department of Defense in the US or the Ministry of Culture, Science, Research and Technology in Germany (Hemmert 1995). Nevertheless, certain ministries are predominantly active in certain areas. Concerning industrial technology, the main policy actor is MITI and therefore, the subsequent analysis will focus on MITI's activities in this field.

It is noteworthy that MITI's share of the total governmental R&D budget is only around 12 percent, compared with MESC's share of more than 45 percent and STA's share of over 25 percent (Kagaku Gijutsucho 1996). This indicates again the low level of governmental expenditures for industrial R&D in Japan.

MITI's technology policy aims at supporting industry's activities in the field of R&D. Its focus in the preceding decades was largely on the identification and selection of strategic areas with high innovative potential, and the formulation and publication of visions which addressed these strategic areas. Moreover, national R&D programs have been initiated in order to stimulate industrial R&D activities. This rather indirect policy approach allowed MITI to play an effective role in inducing R&D activities within the private sector with a rather limited budget. The technology policy has to be seen in the context of MITI's industrial policy as a whole with its emphasis on influencing industrial companies' strategy formulation by envisioning future directions.[2]

Japan has adopted flexible industrial policies throughout its economic development, all of which reflect the international, natural, social, cultural and historical environment of the postwar period (Watanabe 1990). In the late 1940s and 1950s, Japan made every effort to reconstruct its war-ravaged economy, laying the foundation for viable economic growth by introducing a 'priority production system' which allocated limited raw materials, capital and foreign exchange to strategic industries. Through these measures, the consolidation of the economic base and a high increase in industrial productivity was achieved. During the 1960s, Japan began to open its economy to foreign competition by liberalizing trade and the flow of international capital. Thereby, it achieved rapid economic growth led by the heavy and chemical industries. On the other hand, the heavy concentration of such highly material-intensive and energy-intensive industries led to serious environmental pollution problems (Watanabe 1973). This necessitated a re-examination of industrial policy which led to a shift toward a knowledge-intensive industrial structure that would reduce the burden on the environment by depending less on energy and raw materials and more on technology (Tsusho Sangyosho 1972). In the 1980s, intensive efforts continued for the strengthening of a more independent system of knowledge creation.

Industrial technology policy initiated by MITI focused on inducing industry to respond to the above historical demands (Watanabe, Santoso and Widayanti 1991). Japan's success over the last four decades in constructing a virtuous cycle between technological development and economic growth in the face of numerous constraints can, together with several other factors mentioned above, partially be attributed to such a dynamic and flexible policy approach corresponding to a dynamic change in the domestic and international environment.

Japan's industrial policy at a turning point

Japan's manufacturing industry, supported by a favorable domestic environment and MITI's technology policy, performed very well in overcoming the energy crises in the 1970s and electronics oriented high-technology development in the first half of the 1980s, when it had to meet vigorous R&D challenges. In line with this path, the source of Japan's leading high technology has been steadily shifting from an imported base to an indigenous base (Tsusho Sangyosho 1988). A new stream of technological innovation suggested that it was necessary to build not only on existing technology, but also to initiate technological innovation which would induce broad new technologies based on new scientific inventions and discoveries, whose results could be used to resolve global problems (Sangyo Kozo Shingikai 1990; Kagaku Gijutsu Kaigi 1992). At the same time, with economic growth and technological advancement, Japan was requested to make a significant contribution to the international community through the R&D process, its outcome and its spill-over effects (Tsusho Sangyosho 1988). Furthermore, confronting economic stagnation and rising concern for future sustainable development due to CO_2 emission resulting from energy use, a new direction was sought recognizing the critical role technology must play in (1) revitalizing the international economy and (2) providing a solution which can simultaneously overcome energy and environmental constraints while maintaining sustainable growth (Watanabe 1993). Identifying such a simultaneous solution was considered the only survival strategy for Japan as it faced crucial energy and environmental constraints.

Under these conditions, Japan's industrial technology programs have reached a crucial point in which the following requests have been made

1 intensified efforts related to basic and creative technology;
2 greater attention to science and technology that provide a solution for simultaneously overcoming energy and environmental constraints while maintaining sustainable growth; and
3 a greater international contribution to innovative R&D and global issues through the R&D process, its outcome and its spill-over effects.

(Kogyo Gijutsuin 1992)

Although MITI had established a policy system which has supported a dynamic technological development, the policy system has been aimed at improving its own effectiveness and does not necessarily take into full account the redundancy of the broader system (Mowery and Rosenberg 1989). In addition, it was primarily oriented to the rapid development and application of industrial technology for commercial use in the marketplace rather than for the accumulation of scientific inventions and discoveries with a view to international contribution. Facing the above-mentioned turning point, MITI's new task became the structuring of a new policy system which encourages forefront efforts in industrial technology to promote R&D on both basic technology and energy and environmental technologies, leading to an increasing trans-national interdependence.

In order to respond to these challenges, MITI decided to consolidate six existing national R&D programs into the following two comprehensive programs in 1993 (Watanabe 1994).

The Industrial Science and Technology Frontier Program

This program entails restructuring the National R&D Program (Large-Scale Project: 1966), the R&D Program on Basic Technologies for Future Industries (1981) and the R&D Program on Medical and Welfare Equipment Technology (1976) by introducing:

- fundamental and creative R&D which will contribute to the further development of economy and society by building a new technology paradigm and thus facilitating technological breakthroughs; and
- mission-oriented R&D to attain the social goal of meeting public demand and a quality of life common to the international standard in advanced countries (Kogyo Gijutsuin 1993).

The New Sunshine Program

Based on the recognition of the two-sided nature of the global environment issue and energy consumption, this program aims at a comprehensive approach for overcoming global energy and environmental constraints while maintaining sustainable growth through the integration of the Sunshine Project (R&D on New Energy Technology, 1974), the Moonlight Project (R&D on Energy Conservation Technology, 1978) and the Global Environment Technology Program (1989) (Sangyo Gijutsu Shingikai 1992).

Moreover, AIST's (the Agency of Industrial Science and Technology of MITI) national research laboratories have been reorganized which included establishing the National Institute for Advanced Interdisciplinary Research, in order to increase the effectiveness of the restructuring of the above-mentioned long-term programs.

MITI's intensive efforts to restructure its long-term national R&D programs by creating both the Industrial Science and Technology Frontier Program and the New Sunshine Program, together with the reorganization of AIST's national research laboratories, have been expected to appropriately meet national demands concerning Japan's industrial technology at a turning point.

Unfortunately, as this expectation has emerged, Japan's industrial technology has had to face the impact of an unexpected paradigm change in the late 1980s brought on by the relaxation of international oil prices (starting in 1983) and the succeeding rise (1987) and fall (1991) of Japan's 'bubble economy.' Looking again at Figure 3.2, we note that the degree of technology substitution for energy has decreased since 1983 followed by a further decline in 1987. Moreover, the complementary relationship between technology and capital has relaxed, transforming into a substitutional one since 1988. The content and the structural background of this paradigm change will be discussed in the next section.

The paradigm change in Japan's industrial technology

The current state of R&D activities in the Japanese manufacturing industry

Figure 3.4 summarizes the average rate of change in R&D expenditure (including expenditure for technology imports) and factors contributing to such change in the Japanese manufacturing industry from 1974–94 over five periods: 1974–8 (after the first energy crisis and before the second energy crisis), 1979–82 (after the second energy crisis and before the fall of international oil prices), 1983–6 (after the fall of international oil prices and before the 'bubble economy'), 1987–90 (during the 'bubble economy'), and 1991–4 (after the bursting of the 'bubble economy'). We can note a significant decrease in R&D expenditure in Japan's manufacturing industry following the bursting of the 'bubble economy.' Statistics published by the Management and Coordination Agency revealed that Japan's manufacturing industry first experienced a decrease in R&D expenditure in 1992, which continued throughout 1993 and 1994. Eventually, a slight increase was measured again in 1995.

The R&D strategies of firms are depicted by their R&D intensity. In

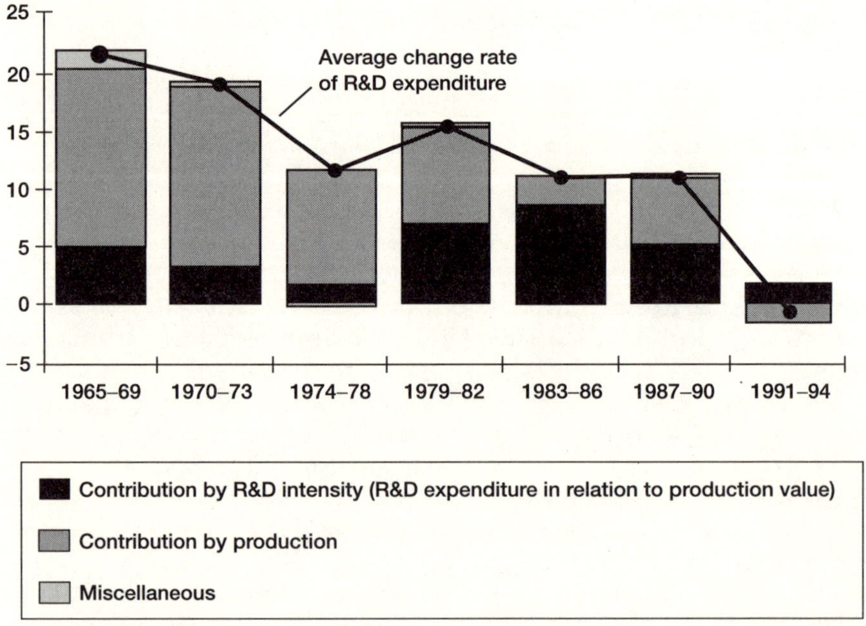

Figure 3.4 Factors contributing to change in R&D expenditures in the Japanese manufacturing industry, 1965–94 (percent)

Source: Authors' calculations.

addition, in order to assess the state of the virtuous cycle between technology and economic development, trends in production (or sales) cannot be overlooked. By identifying contributions of 'inducement by production increase' and 'inducement by strategy' (an increase in R&D intensity), Figure 3.4 also analyzes the contribution of different factors to changes in R&D expenditure in the Japanese manufacturing industry during these five periods. We can note that the increase in R&D expenditure during the period of the 'bubble economy' (10.8 percent per year on average) was largely attributed to 'inducement by production increase' (6.1 percent). After the bursting of the 'bubble economy,' however, the contribution of 'inducement by production increase' fell dramatically (–2.0 percent) and despite a slight rise in the contribution of 'inducement by R&D strategy' (1.3 percent), R&D expenditure decreased (–0.7 percent). Similar trends can be observed in major sectors of the manufacturing industry.

This demonstrates the danger of the virtuous cycle's possible collapse by the following chain reaction: a decrease in 'inducement by R&D strategy' in the period of the 'bubble economy' → stagnation of technology in the period of the bursting of the 'bubble economy' → a decrease in the contribution of technology to an increase in production → a decrease in the 'inducement by production increase' → stagnation of R&D expenditure.

Structural background of the stagnation

Figure 3.5 summarizes the trends in investment objectives within the Japanese manufacturing industry from 1986–95. We can note that the R&D investment's share of total investment decreased substantially (from 12.9 percent in 1986 to 10.5 percent in 1990 and to 8.9 percent in 1994) with its peak in 1987 (13.2 percent). At a time when the nominal R&D expenditures of Japanese manufacturers were still rapidly increasing, their share of overall investment was already shrinking. This change is believed to be a result of a decision made during the period of the 'bubble economy' to focus investment on increasing production capacity by means of non-innovative investment (this share increased from 22.8 percent in 1986 to 32.5 percent in 1991) rather than R&D investment which required consistent innovative efforts with high-risks (Iida 1993).

Another factor leading to a decrease in R&D intensity (stagnating inducement by R&D strategy) was an illusion caused by the differences between nominal and real prices (Watanabe 1992). Firm strategy is generally developed by examining current data and trends based on these data. This behavior can be equally applied to efforts to increase the R&D intensity, which is considered to be a symbolic indicator for investment in the future. This indicator is generally measured by current data (a ratio calculated in current prices). Because this indicator is a ratio, there was an illusion that the ratio represented real R&D efforts. However, the ratio

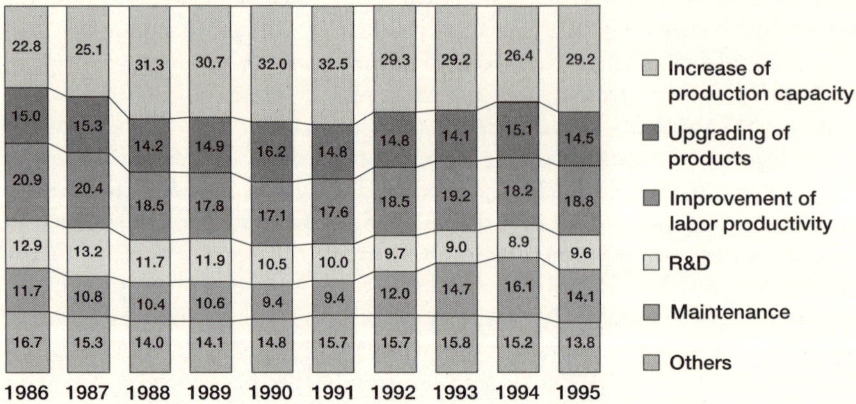

Figure 3.5 Investment objectives in the Japanese manufacturing industry, 1986–95 (percent)

Source: Nihon Kaihatsu Ginko Chosabu (Japan Development Bank, Research Department: various issues).

does not necessarily represent real R&D efforts due to a discrepancy of deflators between R&D expenditures and sales. The R&D deflator reflects the structure of R&D expenditures and contains corresponding shares of land and buildings, labor (researchers), materials, machinery, instruments and equipment.

Due to a sharp increase in the prices of land and buildings as well as labor during the period of the 'bubble economy,' the R&D deflator exceeded the sales deflator, as illustrated in Figure 3.6, causing R&D intensity measured in current prices to be higher than the ratio measured in constant prices (which represents the real efforts to carry out R&D).

Table 3.2 compares trends in R&D intensity measured by both current prices (nominal) and constant prices (real) in Japan's total manufacturing industry and major sectors over the period between 1976 and 1990. It shows that R&D intensity in real terms stagnated or decreased during the 'bubble economy,' while R&D intensity in nominal terms increased, thereby giving rise to the above illusion.

In addition to the above techno-economic factors, we should not overlook two more tendencies concerning Japan's industrial technology itself. Economic calculations have revealed that the annual rate of obsolescence in Japanese manufacturing technology has significantly increased between 1970 and 1990. In order to respond to such increasing obsolescence, industry's R&D efforts concerning new technology became shorter with respect to the time-lag to commercialization. While innovative R&D generally requires a relatively longer time-lag to commercialization, this trend indicates non-innovative R&D. The cyclical trends typically observed in the current stagnation of high-technology industries such as electrical machinery and

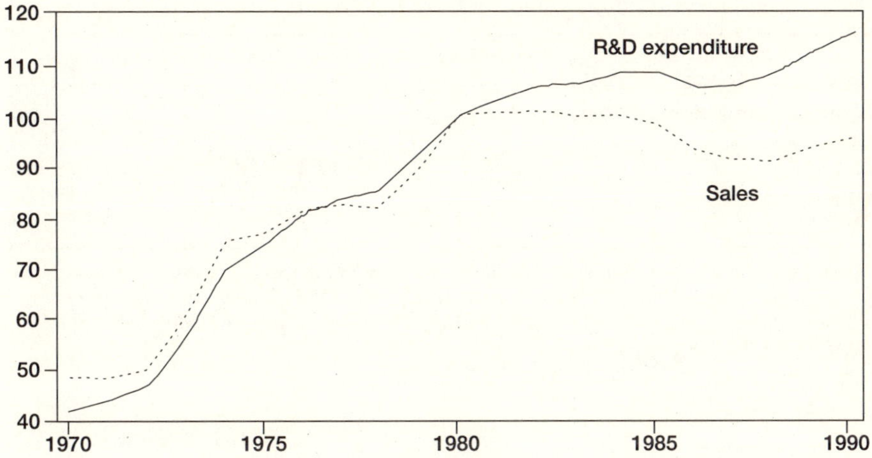

Figure 3.6 Trends in the deflators of R&D expenditure and of sales in the Japanese manufacturing industry, 1979–90

Source: Calculations based on data from Somucho Tokeikyoku (Statistics Bureau, Management and Coordination Agency) (1996 and various issues); Kagaku Gijutsucho (Science and Technology Agency) (1996 and various issues); Nihon Ginko (Bank of Japan) (1995 and various issues).

Note: 1980 = 100, using 1980 constant prices.

transport equipment resulting from demand stagnation can be attributed to such a stagnation of innovative products.

Summarizing this section it has been shown that, during the bubble period of the second half of the 1980s, while nominal R&D expenditures in the Japanese manufacturing industry were still rising, their real amount was already stagnating or even shrinking, as management's attention had been drawn away from R&D to expansion of production capacity. Moreover, there are signs that industrial R&D has become less innovative, as lead times from development to production have become shorter. Noteworthy is the fact that the complementary relationship between technology and capital that was sustained throughout decades, turned into a substitutional one during that period.

As a result, the Japanese industry has been affected by a double punch after the bursting of the bubble. First, R&D expenditures must be reduced as a consequence of depressed business conditions. Second, because of stagnating R&D efforts during the bubble, firms are lacking new innovative products that should have been helping them to improve their business performance. In other words, the long-lasting virtuous circle between capital investment and technological advance appears to be on the verge of collapse.

Table 3.2 Trend of R&D intensity in major sectors of the Japanese manufacturing industry, 1976–90 (annual growth rates in current prices)

	1976–82		*1983–86*		*1987–90*	
Manufacturing industry total	4.4	(4.0)	9.1	(7.1)	2.7	(0.8)
Chemicals	3.2	(4.2)	9.1	(5.3)	3.2	(1.4)
Iron and steel	8.9	(5.7)	15.1	(12.7)	−1.9	(−2.6)
Ceramics	6.4	(7.1)	15.5	(14.6)	−2.4	(−4.0)
Machinery	4.7	(0.1)	5.3	(3.9)	2.5	(−1.7)

Sources: Calculations based on data from Somucho Tokeikyoku (Statistics Bureau, Management and Coordination Agency) (1996 and various issues); Kagaku Gijutsucho (Science and Technology Agency) (1996 and various issues); Nihon Ginko (Bank of Japan) (1995 and various issues).
Note: Figures in brackets are in constant 1985 prices.

Summary and implications: Japan's technological innovation system at the crossroads

In the second and third sections of this paper, a virtuous interaction between vigorous R&D efforts and economic development has been analyzed. Supported by such an interaction, the sustainability of the system, particularly against crises and external shocks, was maintained. However, as reviewed in the fourth section, the current state of R&D stagnation hints at the collapse of this interaction.

In the years after the bursting of the bubble, both capital investment and R&D expenditures in the Japanese manufacturing industry have been shrinking. Currently, they appear to be increasing once again, but their pace of expansion is rather slow. The strong positive interaction between them appears to have weakened, to say the least. Nevertheless, it should not be forgotten that these two indicators do not represent the whole technological innovation system of a country. As has been shown in the third section, Japan's excellent performance in the postwar decades had also been supported fundamentally by economic, socio-cultural, and policy factors. Basically, this favorable environment is still in place, providing the country's industry with good opportunities to recover from the stagnation of the 1990s. Also, it has to be kept in mind that the rapid expansion of the Japanese economy in the postwar decades was essentially a catch-up process to the West. As both the GDP per capita and the amount of R&D expenditures have now reached the level of the other leading economies in the world, a slowdown of growth rates in itself does not have to be perceived as a catastrophe.

The experience of the bubble and its aftermath have clearly shown, however, that there is no everlasting automatism that guarantees a preservation of the virtuous circle between technology and economy. Rather, reinforced R&D efforts are urgently needed to revive the dynamism that drove the

Japanese economy until the 1980s. The renewed increase of industrial R&D spending since 1995[3] suggests that Japanese manufacturing firms, notwithstanding their still constrained business conditions, see this necessity to invest into the future. Therefore, on the threshold of the twenty-first century, there seems to exist at least some possibility to reconstruct the virtuous cycle of R&D and capital investment that has supported the Japanese economy in former decades.

NOTES

1 Noteworthy, the main reason for the further increase in the 1990s was the sharp evaluation of the Yen against the other major currencies. Measured by Purchasing Power Parities, Japan's share of the world's GDP has been stagnating in recent years.
2 This does neither mean that MITI was always correct with its long-term visions, nor does it imply that the Ministry and industrial companies have in all cases been of the same opinion concerning strategy issues. A famous counterexample is that of the automobile industry, when car makers in the 1960s successfully opposed MITI pressure to merge themselves (Johnson 1986).
3 For a detailed description of the recent development of industrial R&D spending, refer to Chapter 8 of this volume.

REFERENCES

BMBF – Bundesministerium für Bildung, Wissenschaft, Forschung und Technologie (1996) *Bundesbericht Forschung 1996*. Bonn: Bundesministerium für Bildung, Wissenschaft, Forschung und Technologie.

Englander, A. and A. Mittelstädt (1988) Total Factor Productivity: Macroeconomic and Structural Aspect of Slowdown. In *OECD Economic Studies,* no. 10, Spring 1988, pp. 7–56.

Goto, A. (1993) *Nihon no gijutsu kakushin to sangyō soshiki* (Innovation and industrial organization in Japan). Tokyo: Tokyo Daigaku Shuppankai.

Hemmert, M. (1995) *Nihon no kagaku gijutsu seisaku no tokuchō* (Features of Japanese science and technology policy). Tokyo: unpublished.

Iida, T. (1993) *Baburu hōkai wa nihon botsuraku no hajimari. Gijutsu kakushin o okotatta tsuke wa ōkii* (Burst of the bubble economy signals fall of Japan, big bill for neglect of technological innovation duty). In *Shūkan Ekonomisuto*, 6 April 1993.

Johnson, C. (1986) *MITI and the Japanese Miracle. The Growth of Industrial Policy, 1925–1975*. Tokyo: Tuttle.

Jorgenson, D.W. (1995) *Productivity. Volume 2: International Comparisons of Economic Growth*. Cambridge and London: The MIT Press.

Kagaku Gijutsu Kaigi (Council on Science and Technology) (1992) *Kagaku gijutsu no sōgōteki kihon hōsaku* (Science and Technology Basic Policy). Tokyo: Kagaku Gijutsu Kaigi.

Kagaku Gijutsucho (Science and Technology Agency) (1996) *Kagaku gijutsu hakusho* (White Paper on Science and Technology). Tokyo: Okurasho Insatsukyoku.

Keizai Kikakucho (Economic Planning Agency) (1991) *Chōki sakkyū shūyō keiretsu kokumin keizai keisan hōkuku* (Report on national accounts from 1955 to 1989). Tokyo: Okurasho Insatsukyoku.

Kogyo Gijutsuin (Agency of Industrial Science and Technology) (1992) *Tenkanki no R&D. Sentā obu ekuserensu no tamenteki jōsei* (R&D at a turning point. The all-around fostering of Centers of Excellence). Tokyo: Tsusho Sangyo Chosakai.

—— (1993) *Sangyō kagaku gijutsu kenkyū kaihatsu seido* (Industrial Science and Technology Frontier Program). Tokyo: Kogyo Gijutsuin.

Mowery, D.C. and N. Rosenberg (1989) *Technology and Pursuit of Economic Growth*. New York: Cambridge University Press.

Nihon Ginko (Bank of Japan) (1996) *Heisei 7-nenpan keizai tōkei nenpō* (Economic Statistics Annual 1995). Tokyo: Nihon Shin'yo Chosa (annual issues).

Nihon Kaihatsu Ginko (Japan Development Bank) (1984) Henbō suru kenkyū kaihatsu tōshi to setsubi tōshi (Changing R&D investment and capital investment). In *Setsubi Toshi Kenkyu '84*, pp. 3–53.

Nihon Kaihatsu Ginko Chosabu (Japan Development Bank, Research Department) *Chōsa* (various issues). Tokyo: Nihon Kaihatsu Ginko Chosabu.

Odagiri, H. (1994) *Growth through Competition – Competition through Growth: Strategic Management and the Economy in Japan*. Oxford: Clarendon Press.

Odagiri, H. and A. Goto (1993) The Japanese System of Innovation: Past, Present, and Future. In Nelson, R.R. (ed.) *National Innovation Systems. A Comparative Analysis*. New York and Oxford: Oxford University Press, pp. 76–114.

OECD – Organization for Economic Co-operation and Development (1991) *Energy Balances of OECD Countries*. Paris: Organization for Economic Co-operation and Development.

Sangyo Gijutsu Shingikai (Industrial Technology Council) (1992) *Sangyō kagaku gijutsu bun'ya ni oite kongo kenkyū kaihatsu ga kitai sareru kadai* (R&D subjects expected to be essential in the field of industrial science and technology). Tokyo: Sangyo Gijutsu Shingikai.

Sangyo Kozo Shingikai (Industrial Structure Council) (1990) *90nendai no tsūsan seisaku bijon* (Vision for industrial policy in the 1990s). Tokyo: Tsusho Sangyo Chosakai.

Somucho Tokeikyoku (Statistics Bureau, Management and Coordination Agency) (1996) *Heisei 7-nen kagaku gijutsu kenkyū chōsa hōkoku* (Report on the Survey of Research and Development). Tokyo: Nihon Tokei Kyokai (annual issues).

Tsusho Sangyosho (MITI) (1972) Ekologī to sono sangyō seisaku e no ōyō (Ecology and application of its concept to industry policy). In *Tsūsan Jānaru*, vol. 5, no. 2, pp. 63–88.

—— (1988) *Sangyō gijutsu hakusho. Sangyō gijutsu no dōkō to kadai* (White Paper on Industrial Technology: Trends and Future Tasks in Industrial Technology). Tokyo: Tsusho Sangyosho.

United Nations (1996) *National Accounts Statistics 1996. Analysis of Main Aggregates*. New York: United Nations Statistical Office (annual issues).

US Department of Commerce (1990) *Japan as a Scientific and Technological Superpower*. Washington, DC: US Government Printing Office.

Wakasugi, R. (1986) *Gijutsu kakushin to kenkyū kaihatsu no keizai bunseki* (Economic analysis of technological innovation and R&D). Tokyo: Toyo Keizai Shinposha.

Watanabe, C. (1973) Nihon keizai no ekologiteki bunseki (An ecological analysis of the Japanese economy). *Keizai Seminā*, no. 211, January 1973, pp. 29–43.

—— (1990) Japanese Industrial Development. In *Australian Journal of Public Administration*, vol. 49, pp. 288–94.

—— (1992) Seizōgyō no kenkyū kaihatsuhi baburu ki ni jijitsu genshō hajimaru (R&D intensity in the manufacturing industry started to decrease during the bubble). In *Nihon Keizai Shinbun*, 25 November 1992.

—— (1993) Energy and Environmental Technologies in Sustainable Development, A View from Japan. In *The Bridge*, vol. 23, no. 2, pp. 8–15.

—— (1994) *Japanese Industrial Science & Technology Policy at a Turning Point – MITI's Role and Its New Initiative*. Paper presented to International Conference on Understanding Government R&D Investment Decisions, Washington, DC.

—— (1995a) *The Interaction between Technology and Economy: National Strategies for Constrained Economic Environments – The Case of Japan 1955–1992*. Laxenburg, Austria: IIASA Working Paper 95–16.

—— (1995b) The Feedback Loop between Technology and Economic Development: An Examination of Japanese Industry. In *Technological Forecasting and Social Change*, vol. 49, pp. 127–45.

Watanabe, C. and T. Clark (1991) Inducing Technological Innovation in Japan. In *Journal of Scientific & Industrial Research*, vol. 50, pp. 771–85.

Watanabe, C. and Y. Honda (1992) Inducing Power of Japanese Technological Innovation, Mechanism of Japan's Industrial Science and Technology Policy. In *Japan and the World Economy*, vol. 3, 357–90.

Watanabe, C., I. Santoso and T. Widayanti (1991) *The Inducing Power of Japanese Technological Innovation*. London: Pinter Publishers.

Yoshikai, M. (1985) *Nihon no sangyō gijutsu seisaku* (Japanese Industrial Technology Policy). Tokyo: Toyo Keizai Shinposha.

Part III

ORGANIZATION, MANAGEMENT AND BUSINESS–GOVERNMENT RELATIONS

4

HIGHER EDUCATION IN JAPAN FROM THE PERSPECTIVE OF R&D

Shinichi Yamamoto

Introduction

Today, every advanced country has its own science system, by which I mean the combination of various elements essential for performing scientific research and development. The higher education system is one of the most important elements for the science system in each country. However, there are wide international differences in its structure. Among advanced countries, Japan is known for its huge volume of higher education. In 1995, there were 1,161 universities and colleges, and total enrollment was about 2,820,000 at the undergraduate and junior college levels combined. Of the 18-year-old population 46 percent advanced to higher education in that year (Monbusho 1996b). Graduate students, however, were fewer in number compared with the huge number of undergraduate students. There were 110,000 students in masters degree programs and 44,000 in doctoral degree programs.

It has been widely said that the Japanese education system has played a pivotal role in the development of Japanese economy and industry and, later, of the innovation system. Regarding higher education, universities have been training future engineers and researchers who work for industry, and they have also been creating and introducing new knowledge and technology from Western countries which has been transferred to the industrial sector for production. Universities and colleges have recently accounted for about 20 percent of the total Japanese R&D expenditure (Monbusho 1996b).

Furthermore, Japanese higher education has the important function of identifying highly qualified manpower through its demanding entrance examination system, in which young people must demonstrate what they achieved in their high school days. This function has caused various educational problems including the credential problem, which places too much emphasis on people's academic background. However, people have accepted the usefulness of higher education for their later career and this notion has greatly motivated people to advance to higher education.

Thus, universities and colleges, particularly the former ones, have had these three main functions:

1 training researchers and engineers as well as other types of professional manpower;
2 performing research and development; and
3 identifying highly qualified young people who should play an important role in Japanese society.

These functions, however, are now changing because of the massive expansion of higher education and also the increasing sophistication of science and technology. The research and research training functions in universities are said to need re-evaluation and improvement, while the students' various educational demands must be satisfied by this mass higher education system.

In this chapter, I will mainly consider the recent effort to improve the research infrastructure under the expansion of higher education. First, however, a brief history of the higher education system since the Second World War is given in the next section to enhance a proper understanding of the recent trends.

Brief history of the Japanese higher education system after the Second World War

It may be a coincidence that the Japanese higher education system has witnessed a structural change every fifteen years since the Second World War (Table 4.1). The first stage lasted until 1960. In this period, most of the efforts were devoted to the reconstruction of the higher education infrastructure which was destroyed by the Second World War and also to the organization of a new education system. This became larger than the prewar system by merging non-universities with universities. People's main concern was how to maintain the quality and the dignity of prewar universities, which were supported by the notion of 'university autonomy' and 'freedom of research.' The enrollment ratio of the 18-year-old population still remained around 10 percent. Higher education at that time was based on the so-called 'elite higher education system', wherein academic research and training of the future elite were the main roles of universities and colleges.

The second stage began around 1960, when the Japanese economy entered its period of high growth. With economic growth, industry's demand for qualified manpower in science and engineering increased rapidly, and policy makers expanded higher education in these areas. At the same time, higher education became affordable for many more people than ever before owing to the growth of household income. People's desire for higher education pushed the higher education system to expand mainly in the social sciences and

Table 4.1 Brief history of Japanese higher education after the Second World War

	Features	*Enrollment ratio (%)*
1945–60	**New university system after the Second World War** • reconstruction of infrastructure • mergers of various pre-war schools of higher education schools with universities • university autonomy	10
1960–75	**Quantitative expansion** • rapid expansion • manpower policy • role of universities under dispute	10–38
1975–90	**Higher education planning** • control of expansion • geographical dispersion • new types of higher education	35
Since 1990	**Diversification of higher education** • deregulation as part of the university reform • selective allocation of resources for research • teaching and/or research	35–46

Source: Arranged from Yamamoto (1996: 14).

humanities. The enrollment ratio jumped up from 10 percent to 38 percent during only fifteen years between 1960 and 1975 (Monbusho 1996a: 75). This rapid quantitative expansion of enrollment did not allow the maintenance of the traditional 'elite' higher education system.

Thus, a large disjunction developed between the majority of the new students, who were little concerned about the traditional norm of 'elite' higher education, and the higher education system itself; serious trouble ensued at universities and colleges in the late 1960s. From such troublesome experiences, a new experiment emerged in the mid-1970s. This was the introduction of Higher Education Planning, aimed at qualitative improvement of higher education which was badly influenced by the quantitative expansion at that time.

Between 1975 and 1990, the third postwar stage of the higher education system in Japan, many measures were introduced to improve the system. Some of them were successful and others were not. In every case, the Government, represented by Monbusho (Ministry of Education, Science, Sports and Culture, MESC) took a strong initiative for the reform of higher education. The most drastic measure was to limit the foundation of new universities and colleges and to strongly encourage them to locate themselves outside Tokyo and other urban areas. By this measure, the enrollment ratio was forced to remain at around 35 percent during the first ten years of this period. Several new types of higher education institutions were also set up during that time. The university of Tsukuba, a well-known representative of the new type of university, started its operation in 1975.

Approaching the 1990s, again a growing population of 18-year-olds started to enter higher education. Monbusho responded to this new growth in students by changing its policy from a regulation-oriented to a deregulation mode. This deregulation enabled universities and colleges to expand again and the enrollment ratio rose to 46 percent in 1995. Deregulation has also influenced the curriculum at institutions of higher education, leading especially to a decline of liberal arts subjects and the growth of professional/practical subjects. Universities and colleges are functioning no longer in the way that they did during the 'elite' system period of the 1950s and 1960s, but have become an integral part of the 'mass' higher education system.

Furthermore, a sharp decline of the 18-year-old population from the present until the decade after 2010 is projected. The ratio of decline will be nearly 40 percent, i.e. from 2 million in the early 1990s toward 1.2 million in 2010 (Monbusho 1996a: 75). Each institution is now seeking its own way to survive in this austere environment by responding more to the consumer-oriented students' needs, such as for practical and vocational training useful for their future career.

At the same time, the problem emerged of maintaining and promoting research and development at universities, which used to be the core of their function. How could advanced research and development be maintained and promoted in the face of the strong expansionist trend of higher education, and also in light of the limited resources? This has become a central issue for policy-makers as well as for each university.

The relation between research and teaching

The relation of research to teaching has been under dispute within as well as without academia. The happy era of the Humboldtian ideal of a close relationship between research and teaching, established and upheld throughout the nineteenth century, has been uprooted by the advance of mass higher education. The expansion has especially influenced teaching contents, which should primarily respond to various social needs. Students are no longer necessarily to be the elite in various sectors of society. Thus, they have come to require more practical and vocational-oriented services from university teaching. Research, on the other hand, has become more and more sophisticated and university faculties must compete with each other for more advanced research not only within the domestic market but also in international settings.

Faculties in the university sector, generally speaking, have been increasingly expected to respond to the changing teaching demands of students and society. According to a comparative international study conducted by the Carnegie Commission on Higher Education, however, faculties in Japan exhibited a strong preference for research over teaching. Over 70 percent of Japanese faculty members said that their first priority activity was research, not teaching, while only 37 percent of their American counterparts agreed to this view (Boyer 1994: 81). This fact explains much about why Japanese professors are so eager to expand their graduate schools and research infrastructure compared with their weak enthusiasm for better teaching.

Expanding graduate education is an approach to secure research activities within the higher education system because undergraduate teaching is no longer able to correspond to the recent advanced research. Indeed, in Japan, the growth of graduate education is amazingly rapid. Although the enrollment of graduate education is still small, the annual growth rate was about 7 percent between 1955 and 1995 at both masters' programs and doctoral programs, compared with 4 percent for undergraduate programs during the same period (Yamamoto 1996: 7). The growth rate of graduate education is also higher than those of other major countries. This high rate is attributed not only to emerging social needs for graduate students but also to a growing desire by university faculties to shield their research environment from the rising tide of expanding undergraduate programs.

Compared with the general trend, however, the growth of master's programs in engineering since the 1970s is exceptional, in that it has been supported by the real demand for graduate students to be future engineers in industry. On the other hand, in other fields, especially in doctoral programs, the expansion of graduate education does not correspond to growing needs for graduates, i.e. for PhDs and other degree holders. There might be an actual overproduction problem of PhDs. Nowadays, all national universities and more than half of private universities have graduate programs in their systems. However, as shown in Figure 4.1, the situation is quite different from field to field. In the humanities and social sciences, the prospects for future growth are still vague.

It seems almost impossible however, to transform all existing universities and colleges into 'research universities' in an American sense. Even in the US, only a small portion of the institutions in the higher education system can be research-intensive. The rest are expected to respond to various educational needs other than advanced research. The direction of policy is, thus, to select relatively few and, by concentrating resources on them, to protect and promote advanced research and research training against the expansion of the higher education system. This direction might be explained by the concentration of resources. Despite the growth of graduate education as a whole, graduate education in doctoral programs, for example, is heavily concentrated in a few research intensive universities like Tokyo University and Kyoto University. As mentioned above, however, since the beginning of the 1990s, the higher education system has entered its fourth postwar stage and

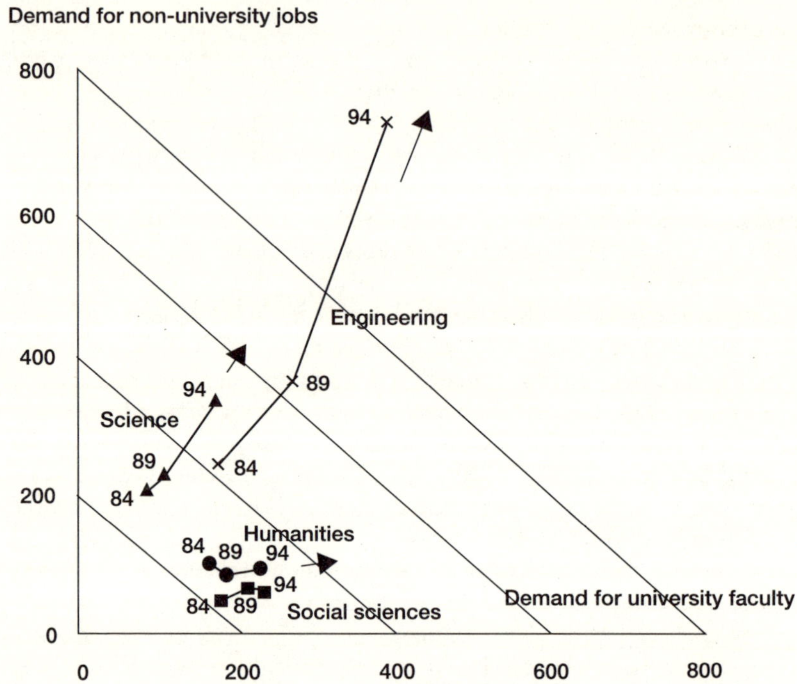

Figure 4.1 Demand for graduate students of different disciplines, 1984–94 (absolute numbers)
Source: Yamamoto (1996: 232).

the key terms are 'diversification,' and 'selective allocation of limited research resources.'

Improvement of research bases and organizational structures

Despite the strong need for advanced research in some universities, the university research environment worsened in the 1980s. Owing to Japan's governmental budget deficit problem from the early 1980s, allocations for national universities have been almost frozen. Research laboratories at universities suffered from research funding shortages and from outdated facilities. National universities had depended mainly on general university funds allocated by Monbusho according to a standard formula. They might be losing their appeal to graduate students and young post-docs because private universities, governmental research institutes and research laboratories in private industry tend to attach increasing importance to basic research in the late 1980s and they invested large amounts of funds in their in-house research. Thus, how to attract young people to science and engineering research at university laboratories became one of the biggest policy issues in higher education policy just at the beginning of the 1990s.

In response to this, Monbusho started to implement several policies for improving the research environment of universities. Typical were the endowment of special funds for graduate schools (institutional base) and new fellowships for doctoral students and post-doctoral researchers (individual base). The aim of these new policies, in addition to the growing amount of competitive grant-in-aid programs, was to devote more resources not to institutions and students but to certain selected institutions and researchers whose research quality and performance were above par. Besides such policies, Monbusho has encouraged universities to foster joint research between universities and industry and other types of interactions between both sectors.

The grant-in-aid programs are research grants provided with the aim of encouraging able researchers in universities to do good and innovative research. These grants are given after a competitive evaluation process for each proposal. In prior decades, the amount of the grant-in-aid was only a small portion of the total research budget and thus it was supplementary. In recent years, however, Monbusho's continuous effort to increase the grant-in-aid program has lifted the budget to 100 billion Yen (Monbusho 1996c) which is almost equal to the research portion of general university funds. So, nowadays, the grant-in-aid program is indispensable for active researchers in order to survive in the competitive research environment.

Research cooperation with industry is another scheme for encouraging research activities. Since the mid-1980s, Monbusho has encouraged universities to engage in such cooperative projects by several measures. Joint research between national universities and private companies, for example, is cooperative research on themes of common interest to the parties involved. Universities accept researchers and research funds from private companies for carrying out research. By combining a university's research capability with a company's technology, this joint research system is expected to produce better results than separate individual research projects in each sector. In fiscal year 1995, 1,704 joint-research projects were carried out at the expense of 18 billion Yen. In addition to several measures of cooperative research, universities accept grants and endowments from industry. National universities accepted 49 billion Yen in fiscal year 1995 (Monbusho 1996c).

As a result, the funding structure for university research has been changing. Universities and their faculties now increasingly expect extra funding from private companies or competitive governmental grant-in-aid programs other than general university funds (see Table 4.2). At some laboratories within research intensive universities, for example, in engineering, more than one third of research funds come from the private sector. As a result, these external funds have become essential for the maintenance of the laboratories.

Besides these measures, national universities received supplementary

Table 4.2 The growth of major university research funding (1985 = 100)

	1985	*1990*	*1995*
General university fund	100	119	149
Grant-in-aid from Monbusho	100	133	220
Grant and donations from industry	100	175	236

Source: Calculated from Monbusho (1996d and various years).

budgets from Monbusho in recent years as the result of governmental policy of public investment to stimulate the depressed Japanese economy. In 1996, another new governmental policy was launched (Monbusho 1996c: 322). Several large funds have been newly created for the purpose of funneling investment to some excellent large research projects. This system will open the way for universities to have 'multi-funding sources,' drawing not only from Monbusho, but also from other government agencies.

Now, as the problem of research funding shortage seems to have decreased in importance, the next step is to improve the system and structure of research at universities. That means to fine-tune the university research structure to appropriately spend the much-improved research funds. The lack of support staff, both for direct research activity and for research administration, is very serious. Red tape has prohibited professors from paying support staff or graduate students salaries from their research grant and any sources other than the regular governmental budget for salary. This is only one example. The Science and Technology Basic Plan makes various suggestions for the improvement of research institutions. The most important task is how to bring this plan into effect.

Concluding remarks: the change of orientation of the higher education system

While struggling to retain the research function as an element of the education system as discussed above, the system itself will be, or should be, changed toward strengthening the future research function of universities and other higher education institutions.

To date, Japanese higher education has been aimed mainly at the young, because one of the primary functions of higher education has been identification of the most able young. This function as gatekeeper has prevented adult students from studying at universities, since education there has brought them scant benefit. Most of the students have attended universities on a full-time basis and ceased all study at their graduation from undergraduate programs. Selection of the elite young has occurred exclusively at the stage of graduation from the undergraduate level. This selective system caused serious competition for the undergraduate entrance examination. Thus, after graduation, people are evaluated mostly by 'from which *university* did you graduate?' rather than 'what did you *learn* at your university?' In spite of the diversification of students' needs for education and training, most faculty members have cleaved very conservatively to their own interests, and their teaching style has been heavily discipline-oriented so as to prepare future academics.

However, this situation is bound to change in the future. The upcoming key concept is 'lifelong learning,' and this is closely tied with the coming 'knowledge-based society.' As shown in Table 4.3, higher education in Japan is changing its orientation from the traditional one toward a lifelong learning orientation. This will, of course, affect not only educational activities but also the research function of higher education institutions.

Table 4.3 From traditional learning mode toward lifelong learning mode

Traditional mode	*Lifelong learning mode*
Only for the young	Also for adults
Full-time students	Full-time + Part-time
Undergraduate centered	Undergraduate + Graduate
Selection of the young	Training for adults
Competitive access	Open access
Discipline oriented	Problem oriented
Academic training	Professional training
What university did you graduate from?	What did you learn at your university?

Source: Author's composition.

REFERENCES

Boyer, Ernest L. (1994) *The Academic Profession. An International Perspective*. New York: The Carnegie Foundation.

Monbusho (1996a) *Japanese Government Policies in Education, Science, Sports and Culture 1995*. Tokyo: Publication Bureau of the Ministry of Finance.

—— (1996b) *Statistical Abstract of Education, Science, Sports and Culture 1996*. Tokyo: Publication Bureau of the Ministry of Finance.

—— (1996c) *Waga kuni no bunkyo shisaku 1996* (Japanese government policies in education, science, sports and culture 1996). Tokyo: Okurasho Insatsukyoku.

—— (1996d) *Waga kuni no bunkyō yosan 1996* (The budget for education, science, sports and culture in Japan). Tokyo: Monbusho.

Yamamoto, Shinichi (1993) Research and Development versus Traditionalism at Japanese Universities. In *Higher Education Policy*, vol. 6, no. 2, pp. 47–50.

—— (1996) Gakujutsu kenkyū shisutemu kara mita daigaku'in ni kansuru kenkyū (Graduate schools in Japan from the perspective of academic research). In *University Studies*, no.15, pp. 1–287.

5

INTERNATIONALIZING JAPANESE SCIENCE

Brendan Barker

Introduction

A major objective of recent Japanese science policy has been to increase the country's efforts in basic research. As part of this effort, a further priority has been to promote cooperation with the international research community (Arima 1996). Support for basic science is primarily justified on the grounds that it will underpin development of the next generation of technologies needed to sustain economic competitiveness and quality of life. This view is encouraged by international demands that Japan should contribute more to international research activity. These two aims – the promotion of basic science and the promotion of increased internationalization – are mutually reinforcing. Internationalization demands that good science be done in Japan, while good science thrives best as part of an open international research system.

These priorities have been reflected in recent large increases in public-sector science funding and in the aims of the recent Basic Plan for Science and Technology (Samejima 1995). However, improving both the quality of science undertaken in Japan and its attractiveness to researchers in other countries will depend on more than simply increased funding. A number of structural problems have to be addressed if the Japanese research system is to reach its full potential.

A number of Japanese policy-makers have long recognized this and in recent years they have been joined by an increasing number of politicians. From the early 1980s, new ways of supporting and managing research have been investigated. While these have so far been on a small scale, they are significant in the lessons they provide for a more full-scale restructuring of the Japanese research system – for which pressure is mounting.

A particular feature of these initiatives has been an emphasis on increased collaboration at both the national and international level. Increasing collaboration is a world-wide trend reflecting the increasing breadth of knowledge and complexity in equipment and procedures required to undertake research. Indeed, collaborative research is rapidly becoming the dominant mode of scientific production. Increasingly, such collaboration involves researchers from different institutions and from different countries. However, while Japan is no exception to this trend, levels of inter-sectoral and international collaboration remain lower than for most other industrialized countries.

Globally, the internationalization of R&D activities – in which international collaboration is an important element – has expanded greatly in the last decade. The increasing internationalization of science is being driven by many recent developments, including the recognition that the complexity and costs of major research facilities are outstripping the resources of even the most advanced countries, that modern information and communication technologies make remote utilization of major facilities feasible, that increasingly the most serious societal problems (e.g. global environmental change and world population growth) are truly global in scope and scale, and that the number of nations with a leading-edge research capability is increasing rapidly. As discussed elsewhere in this book, this process of internationalization has been led by the private sector. It has also become an important element of much publicly supported research. It is this that I wish to discuss in the subsequent sections.

This chapter looks at Japanese efforts to internationalize the country's science base. First we show that although Japanese science is becoming more global in activity and impact, this is happening only slowly. Partly this is the result of weaknesses in the existing science system – some aspects of which we discuss. We then go on to examine attempts to improve the environment for research in Japan, before looking at specific initiatives intended to promote internationalization in science and technology.

The current status of Japanese science

One of the main justifications given in Japan for supporting basic research is that it will provide the basis for the next generation of technologies. This implies a 'linear model of innovation' in which investment in basic research leads to discoveries which are ultimately used in new products and processes. However, studies have shown that the innovation process is typically more complex than this (although such a model may be a more reasonable description of innovation in some science-based industries). A more robust justification for supporting basic research is that it provides countries with the capability to tap into and exploit new developments no matter how or from where they originate.

While Japan is now second only to the US in the amount it spends on research and development, this still amounts to little over 10 percent of the world total. Furthermore, much of this funding comes from the private sector to support applied research and development used to produce proprietary knowledge. Even with a large increase in support for basic research, the vast bulk of such activity will still be undertaken outside of the country. Access to the results of this research will increasingly require evidence of reciprocal research activity. More importantly, domestic exploitation requires that Japan has the capability – skills and techniques – in place to internalize and utilize the outputs of the world research system, in addition to those of its own (Odagiri and Goto 1997). In this sense basic research can be seen both as the price Japan has to pay to gain access to the pool of world knowledge and as part of the capability-building activity which enables that knowledge to be exploited once acquired. Thus internationalization should not simply be seen as a tool to improve basic science – but as an important justification for doing basic research in the first place.

Given that science is becoming increasingly international in scope, where does Japanese science stand? First, there is some evidence that the amount of Japanese research of international quality is increasing. The number of Japanese papers on the Science Citation Index (SCI) of the Institute for Scientific Information (ISI) has increased from just over 25,000 in 1981 to nearly 37,000 in 1993 (Table 5.1). As a percentage of total world papers published each year this represented an increase from 6.8 percent to 8.8 percent over the same period – Japan is now second only to the US in the number of papers published in the selected journals. If anything, because of the English-language (particularly American) bias of such databases, this underestimates the activity of Japanese researchers. Nevertheless, while far from perfect, analysis of publication patterns does provide a useful indicator of scientific outputs by Japanese researchers.

The quality of research – as measured by citations to publications by Japanese authors – has increased in recent years but remains below that of the US and UK (Table 5.2). However, researchers from China and the NIEs (newly industrializing economies) cite Japanese researchers more often than those from other countries with the exception of the US showing that Japanese research has a particularly strong impact in these countries. Less positively, there is evidence of a more inward looking Japanese research system: although authors in all countries have a strong bias toward citing

Table 5.1 Number of papers by Japanese authors on the Science Citation Index, 1981–93

	1981	*1983*	*1985*	*1987*	*1989*	*1991*	*1993*
Total number of papers	25,086	26,368	29,617	28,894	32,832	34,375	36,674
Percentage of world total	6.80	7.10	7.60	7.60	8.10	8.50	8.80

Source: NSF (1996).

Table 5.2 Selected countries' citations to the international scientific and technical literature (%)

Citing country	*Cited country*								
	US	*UK*	*Germany*	*France*	*Japan*	*China*	*NIE*	*Others*	*Total*
US	69.5	5.7	3.9	2.8	4.2	0.2	0.2	16.5	100
UK	37.7	32.3	4.4	3.5	4.2	0.1	0.3	17.5	100
Germany	37.5	6.6	29.1	3.9	4.8	0.2	0.3	17.6	100
France	39.5	7.0	5.3	25.2	4.9	0.2	0.3	17.6	100
Japan	36.4	5.1	4.2	3.0	38.7	0.2	0.3	12.1	100
China	37.0	5.2	5.9	4.6	8.8	17.9	0.9	19.7	100
East Asian NIEs	42.8	6.5	4.0	3.1	9.1	0.8	19.1	14.6	100

Source: NSF (1996).

domestic sources, this trend is more pronounced in Japan than in any other country, with the exception of the US.

Evidence suggests that Japanese research is stronger in the physical sciences than in the life sciences. Japanese researchers in the physical sciences produce a higher proportion of world papers than those in the life sciences indicating relatively greater 'activity'. Furthermore, physical science papers are typically more highly cited by world's researchers than those in the life sciences – indicating greater 'impact' or quality (Barker 1996).

Publication patterns also indicate that Japanese research activity is becoming more internationalized – albeit slowly. Table 5.3 shows patterns of international collaboration as reflected by internationally co-authored papers. Over the two periods, 1981 to 1987 and 1988 to 1993, the percentage of internationally co-authored papers increased from 7 percent to 11 percent of the total number of papers published by Japanese authors. However, this is still low in international terms.

Growing internationalization is also demonstrated in the increasing mobility of researchers. In 1991 nearly 200,000 researchers and engineers left Japan for other countries, while about 130,000 overseas researchers and engineers came to Japan (NISTEP 1995). However, there is an interesting difference as to where foreign researchers have come from and where Japanese researchers are going to. The majority of scientists and engineers entering Japan have consistently come from other, less developed, Asian countries. On the other hand, most Japanese researchers going abroad go to industrialized countries. Furthermore, most of the scientists and engineers entering Japan appear to do so for training purposes – to acquire knowledge and skills – rather than to contribute knowledge through teaching or original research (NISTEP 1995).

Scientific and political circles in the US and Europe as well as domestically in Japan have sought to redress this imbalance and are urging Japan to

Table 5.3 Patterns of international collaboration in the publication of scientific papers, 1988–93 (1981–87)

		Collaborating country (percent of total international)					
	Total international (%)	*US*	*Japan*	*UK*	*Germany*	*France*	*NIEs*
Source country							
US	14 (9)	na	8 (7)	10 (12)	10 (11)	8 (7)	3 (1)
Japan	11 (7)	43 (35)	na	6 (6)	8 (9)	5 (4)	2 (2)
UK	22 (15)	25 (28)	3 (2)	na	8 (8)	7 (7)	1 (1)
Germany	26 (17)	23 (25)	3 (3)	8 (9)	na	8 (8)	1 (1)
France	28 (19)	21 (23)	2 (2)	9 (8)	10 (10)	na	0 (0)

Source: NSF (1996).

embark on various initiatives to make research in Japan more attractive to foreign researchers (Sigurdson 1995). The next section will briefly describe some of the more significant problems.

Weaknesses in the Japanese research system

The research environment in Japanese universities, graduate schools and government laboratories is, on the whole, poor. Partly, this has been the result of low levels of funding: Table 5.4, below is an international comparison of research funds available per researcher (an average for both natural sciences and humanities). While direct comparison is difficult (because of differences in the way research personnel and funds are defined), it does reflect a real issue.

The relatively low levels of funding for public-sector science have been reflected in a lack of investment in research facilities. As of 1994, about 49 percent of all research facilities in Japanese national universities and about 29 percent of the facilities of government laboratories under the Science and Technology Agency (STA) and the Ministry of International Trade and Industry (MITI) were reported to be more than twenty years old. Likewise, about 82 percent of all R&D equipment in STA and MITI laboratories have passed their expected life (Basic Law 1995).

On average throughout the research system, official figures suggest there is only 0.48 support staff per researcher in Japan, compared to 1.2 in Germany, 1.23 in France, and 1.07 in the UK (Kagaku Gijutsucho 1995). In the university sector the situation is even worse with 0.13 support staff per researcher in Japan, compared to 1.1 in both Germany and France and 1.2 in the UK (Basic Law 1995). As a result, researchers in Japan often find themselves doing mundane laboratory tasks usually undertaken by technicians in other countries.

In the mid-1990s, this resource problem has begun to be addressed. Increased funding for science and technology was a major element of supplementary budgets introduced during 1995. Science and technology also fared well under the budget for 1996, with the government launching a

Table 5.4 R&D expenditure per researcher (million Yen)

	Japan (1994)	*US (1993)*	*Germany (1991)*	*France (1991)*	*UK (1993)*
Government labs	27.00	50.50	26.10	43.70	38.80
Universities	12.00	36.00	17.90	16.60	20.30
Private industry	25.20	29.30	33.40	48.40	29.50
Total average	21.20	31.50	28.30	37.10	28.10

Source: Management and Coordination Agency, quoted in Samejima (1995).
Note: Foreign currencies converted to Yen figures by OECD's Purchasing Power Parities.

series of initiatives to promote basic research, higher education, and innovation. A Science and Technology Basic Law was enacted in November 1995, and the subsequent Science and Technology Basic Plan, approved by the Cabinet in July 1996, urged the government to double the budget for research and development within the next five years. The aim is to raise government investment in research and development to the level of the United States and leading European nations by the turn of the century. While it looks increasingly unlikely that this target will be met, the science budget will probably continue to increase year on year, against an overall decline in public-sector spending.

The authors of the Plan recognize that a simple increase in funding will not be sufficient to improve the quality of science undertaken in Japan. Thus it also suggests specific reforms to add impetus to research activities. Spending is to focus on training, cooperation between research projects and the purchase of equipment and computers. In promoting changes in the way science is organized, the authors argue that 'the biggest priority is to make the environment that surrounds science and technology more flexible, competitive and open' (Kagaku Gijutsu Kaigi 1996).

This will require fundamental changes in the way scientific activity is generally organized and managed in Japan. The science system – particularly in the academic sector – has been characterized as one dominated by small, narrowly focused projects, with funds awarded on the basis of seniority and with a preference for orthodoxy, linked to an institution rather than an individual, and watched over closely by the funding agency (Hayashida 1996).

In the past, in the university sector, research funding has been spread evenly – and far too thinly – among all university research staff. Core funding for research in national universities is allocated in small amounts to every *kōza* (chair or division) of every department. A *kōza* typically consists of a professor, an associate professor and one or two assistants, while research funds are allocated according to a standard formula based on the numbers of students, the nature of the research and other factors. This system has proved very inflexible – particularly in keeping up with rapidly developing research areas. In addition, in many areas of research the *kōza* group does not provide the 'critical mass' required to conduct research effectively, and does not lend itself to the informal and fluid collaborative arrangements which are a feature of many other research systems (Barker 1996).

In the light of attempts to internationalize Japanese science, it is interesting to note that, in the past, many procedures regulating the employment of staff at national research laboratories and national universities made it virtually impossible to hire foreign researchers. This gradually changed during the 1980s when determined attempts were made to attract more foreign researchers. In response to pressure to increase the number of foreign faculty members, Japanese law was changed in 1982 to permit the employment of foreign nationals as faculty members at Japanese national universities.

Monbusho (Ministry of Education, Science and Culture) regulations state: 'in principle, foreign faculty members should be hired on a temporary basis, but, in unavoidable circumstances (the granting of tenure to foreign faculty members) is not illegal' (Geller 1994).

This change in the law has had some effect in that the number of foreign faculty members in national universities increased from a mere handful in 1982 to more than 450 in 1994. However, almost all foreign faculty members of Japanese national universities are still appointed on fixed-term contracts of one to ten years with the possibility, but without any guarantee, of reappointment. In contrast, all Japanese faculty members are tenured (Geller 1994).

The low number of foreign faculty partly reflects a wider problem of academic 'inbreeding'. In 1972 more than 90 percent of the faculty members of Tokyo University were also graduates of that same institution. By 1989 this had fallen only slightly to just under 87 percent. Similarly, more than 80 percent of the faculty members of Kyoto University are also graduates of that university (Geller 1994). Such faculty inbreeding is prevalent at all major universities in Japan and must be reduced before a significant influx of foreigners can be achieved. However, the rules governing faculty appointments make this difficult. Decisions concerning temporary and tenured appointments are made at the level of university or faculty. For this reason Monbusho has no control over individual cases and would be unable to dictate a policy on recruitment – even if it wanted to do so.

These problems are not restricted to the university sector. The Science and Technology Agency highlights low flexibility and competition in public research institutes as a particular problem. The Agency argues that the Japanese R&D structure's main deficiency is the domination of public research laboratories by senior lifetime employees, with the country's respect for seniority still taking precedence over the promotion of the best young scientists (Dawkins 1996). Shizuo Hoshiba, Deputy Director of the group responsible for preparing the 1996 Science and Technology Basic Plan, has argued: 'Japanese society is very conservative and this is especially true in R&D. It is very important to change and to improve researcher's conditions' (quoted in Dawkins 1996).

The need to improve conditions for young researchers is given extra urgency by the fact that STA officials forecast that Japan's 600,000 private and public sector researchers will dwindle to just over 500,000 by 2005 as a result of the country's aging demographic profile and the unpopularity of basic research jobs (Dawkins 1996).

Particular concern has also been expressed over the limited collaboration among researchers across different research fields and across different sectors, i.e. universities, government laboratories and private companies (Samejima 1995). This is a particularly important issue given the low mobility of researchers in Japan. Bureaucratic rivalries have formed a major barrier to public-sector collaboration. For instance, in 1980, at a meeting of the advisory Council for Science and Technology (CST), the head of the STA proposed giving considerable STA money to university professors. Officials from Monbusho reportedly warned academics that if they accepted the money, they would never again give them funds (Koizumi 1993).

There have also been barriers to collaboration between the public and private sectors. For instance, only since 1983 have national university academics been allowed freely to undertake joint research with private companies. Although Monbusho subsequently changed its policy on collaboration, and significant amounts of additional private and public-sector funds for university research are becoming available, levels of collaboration remain low. A collaborative 'culture' is still lacking in much of the university sector: 'Many Japanese companies find it easier to work with a US or UK university than with a Japanese institution' (Dawkins 1996).

Much of Japan's research strength lies in its industrial laboratories – this contrasts with the West where universities and public sector research organizations have played a much bigger role. A number of the larger Japanese companies conduct fundamental research in their central research laboratories. Such research however, is coming under increasing pressure. The collapse of the 'bubble economy' has led both to a decline in private sector R&D funding – the first in over forty years – and a noticeable trend for companies to focus on more near-term research. In the next few years, the onus will be on the public sector to play a bigger role. In the next section we look at efforts to improve the environment for research in the public sector so that it can play this bigger role.

Improving the quality of Japanese science

Increased internationalization depends on high quality research being done in Japan. A number of initiatives have been introduced to improve the research environment in the public sector. For instance Monbusho, accepting the recommendations from a 1992 report by its Science Council, has set itself the task of upgrading Japan's scientific research base to international standards (Gakujutsu Shingikai 1992). To achieve these goals, Monbusho is implementing a range of policies which give priority to the following four areas:

- improving the scientific research environment by increasing research funding and prioritizing allocations, promoting the diversification of sources of funding, and upgrading facilities and equipment;
- promoting the training and securing of researchers through qualitative and quantitative improvement of graduate schools, upgrading and enriching fellowship systems, further promoting researcher mobility and other measures;

- promoting scientific research as a priority by improving joint research systems, forming networks for promoting coordinated cooperative research, developing centers of excellence and other measures; and,
- contributing to international scientific research through support for international exchange of scholars and researchers and joint international research.

Similarly, in its recent White Papers on Science and Technology, the STA has consistently argued that the government should introduce similar measures to improve the country's science base (Kagaku Gijutsucho 1995). Indeed, dissatisfaction with the way science is supported has been a major motivation behind a number of innovative initiatives – beginning with the ERATO (Exploratory Research for Advanced Technology) program in 1981 which is analyzed in detail in another chapter of this book. ERATO is supported by the STA and managed by the Japan Science and Technology Corporation (JST – formerly the Research Development Corporation of Japan – JRDC). In the late 1970s, Gen'ya Chiba, a nuclear scientist who joined JRDC after studying and working in the US, decided that Japan's current system would never allow it to catch up with its global scientific competitors. To tackle this problem, Chiba created a program in which project leaders were given ample funding and a free rein to set up and manage their own laboratories (Normile 1996).

Interestingly, ERATO was one of three programs set up in 1981 advocating *san-gaku-kan* (industry–university–government) joint projects. Of the three programs, however, ERATO was the only one which was fundamentally innovative, and the only one which could be considered a success in promoting inter-sectoral collaboration (Koizumi 1993). One reason for this success is that JST, as a public corporation affiliated with the STA, is not subject to the government's restrictions covering civil service employment. University participation in ERATO has been extensive with about a third of the projects being led by university professors.

The success of the ERATO program – confirmed in two evaluations carried out in 1988 and 1995 – led to the launching of three further programs: the International Joint Research Program (IJRP) in 1989, Precursory Research for Embryonic Science and Technology (PRESTO) in 1991, and most recently the Core Research for Evolutional Science and Technology (CREST) initiative, created in 1995. IJRP is essentially an international version of ERATO. It sets up cooperative projects between Japanese and foreign research organizations based on the principle of equal sharing of costs and facilities. PRESTO provides a limited number of research grants to individual researchers to conduct small-scale exploratory research. The idea behind the latest initiative, CREST, is to give scientists a free hand to pursue a large, ambitious and risky, but potentially high value project in an area of high strategic importance. While CREST projects have a much stronger strategic focus than those funded under ERATO, they do allow the direct funding of university staff. This is in contrast to ERATO where project leaders have not (officially) been allowed to use national university facilities. For many years national university professors heading ERATO projects had to insulate them from their university activities.

The ERATO model is not unique. In 1987 the Institute of Physical and Chemical Research (RIKEN) established its Frontier Research Program (FRP). Like ERATO, FRP was established to provide a flexible framework for conducting long-term leading-edge basic research. It is interesting to note that, like JST, RIKEN is a Public Research Corporation and therefore unencumbered by the regulations covering employment of researchers in the national universities and laboratories. In addition to their emphasis on promoting creative, basic research these initiatives also demonstrate a more flexible approach to recruitment and management, and increased openness to (and indeed encouragement of) participation by foreign researchers.

In addition to these specific initiatives, recent years have also seen significant developments in the wider research system. Three particular trends reflect attempts to improve the research environment: an increasing concentration and selectivity of resources, an increasing promotion of inter-sectoral collaboration, and the increasing use of evaluation.

First, in an attempt to concentrate resources in areas of high quality and/or strategically important research, the Japanese government has recently promoted the idea of 'centers of excellence' (COEs). These were first called for in the General Guidelines for Science and Technology Policy established in 1986, and reiterated in the revision of April 1992. The Guidelines defined COEs as core research institutes with prominent research leaders, up-to-date research information, excellent facilities/equipment, and a substantial research support system.

In 1993 STA began administering a COE scheme encompassing all government research institutes with the exception of those funded by Monbusho. National institutes apply through their managing ministries/agencies to STA and the applications are reviewed by the Prime Minister's Council for Science and Technology (NSF Tokyo Office 1995). STA then provides COE funds to selected institutes to support state-of-the-art basic research which requires collaborative work by many researchers (NSF Tokyo Office 1996).

Monbusho's Science Council had called for the establishment of its own system of COEs in its report of 1992 (NSF Tokyo Office, 1997). After two years of discussion, the plan finally materialized in 1995. Monbusho's COE initiative consists of three elements: the development of new COEs; the advancement of organizations already recognized as COEs; and the improvement of facilities and equipment at targeted institutes to give them the potential to become COEs.

A related trend is that of an increasing diversity in the sources of funding for university research. In a reversal of its earlier reluctance to allow funding from other sources, Monbusho now allows university academics to accept research funds from outside. In 1995, NEDO introduced a new grant scheme. This marked the first time that a government department other than Monbusho had been allowed to award a substantial number of large grants to university researchers. The Science and Technology Agency subsequently announced similar plans (the CREST scheme) to fund research at universities.

The second major trend has been the increasing encouragement of intersectoral collaboration. Thus, for a number of years, Monbusho has tried to improve industry–university cooperation. In 1983, the Ministry established the University–Private Sector Joint Research Program to allow national universities and other institutions to accept researchers and research funding from private industry. Since 1987, Monbusho has commissioned 'Joint Research Centers' at national universities to undertake joint research projects and promote other areas of academic and private sector cooperation. This scheme has not been notably successful. The Ministry is now trying to promote industry–university collaboration by making it administratively easier for universities and colleges to undertake contract research and accept contract researchers.

JST has been more successful in its establishment of high technology consortia in order to develop promising major research results that arise from ERATO, national laboratories, universities, and other sources. JST brings companies together with researchers to explore new applications which hold potential for benefits to the participating companies.

Finally, a relatively recent development has been the increasing use of evaluation. The 1995 White Paper argued that such evaluation should be a high priority, while the Basic Law implicitly recognized this by calling for improvements in the management of R&D funds. Tokyo University was one of the first academic institutions in Japan to undertake an evaluation of some of its research activities. Regular evaluations of research are also undertaken at RIKEN. In addition, the Human Frontier Science Program (HFSP), which is 80 percent funded by the Japanese government, has supported major international evaluations of its activities. An evaluation of ERATO has been undertaken by international teams in 1988 and 1995. STA COEs are subject to an evaluation during the third year of operation.

These developments are indicative of the fundamental change happening to the Japanese science system as it adjusts to the country's shift from a 'catch-up' to 'leading-edge' economy. The new Basic Plan builds on these developments: it proposes an overhaul of the research career system, supports increased intersectoral collaboration, and provides for more support personnel and equipment. If sustained, these developments will lead to a more effective research system. An important component of this system is its increasingly outward looking character. The next section looks at specific ways the Japanese government has attempted to promote increased internationalization.

Promoting the internationalization of Japanese science

In parallel with attempting to improve the quality of Japanese science, the Japanese government has adopted a range of initiatives to promote an increasingly internationalized science and technology system (Georghiou 1993). These include:

- foreign participation in domestic R&D programs;
- encouraging international researcher mobility;
- improving access to the results of Japanese science and technology;
- participating in world-wide activities; and,
- promoting specific collaboration programs.

We will briefly look at each of these areas in turn. Following the lead of the ERATO program, a number of other domestic research programs have been opened to foreign participation. For example, almost 30 percent of researchers and research leaders in RIKEN's Frontier Research Program come from outside Japan. Similarly, there are a large number of foreign researchers working on domestic fusion and environmental research initiatives.

Related initiatives support international researcher mobility. For instance, the International Research Exchange Program began in 1989 as one of several measures in Japan aimed at strengthening international cooperation in its science and technology activities. JST implements the STA Fellowship Program which provides foreign researchers with opportunities to conduct research at Japan's national laboratories and other organizations. MITI's Agency of Industrial Science and Technology (AIST) and the Japan Society for the Promotion of Science (JSPS) support similar fellowship schemes in MITI's laboratories and the universities.

Attempts are also being made to improve access to the results of Japanese science. The Japan Information Center of Science and Technology (JICST) was established in 1957 as the central organization of information activities in the fields of science and technology in Japan. It collects, processes and disseminates scientific and technical information published in and out of Japan. Recently, JICST has been focusing its efforts on distributing Japanese information world-wide. Similarly, the National Center for Science Information Systems (NACSIS), an inter-university research institute under Monbusho, was established in 1986 to gather, organize and provide scholarly information. Like JICST, an increasing amount of its work is taken up with disseminating the results of Japanese research overseas.

An increasing number of foreign scientists have experience of the Japanese research system – a few have even become proficient in the Japanese language. However, language remains a significant barrier to effective international research collaboration between Japanese and foreign researchers. Robert Geller, an American academic at Tokyo University, argues: 'The language problem is real, but it should not be used as an excuse. Foreigners new to Japan can make significant contributions to committees dealing with international issues and they can help in the production of English reports' (Swinbanks 1993).

It is an unfortunate fact of life that while English remains the dominant language for scientific communication the onus will be on Japanese researchers to improve their English skills. This is already happening. There is a general trend for younger researchers to be more proficient at English than more senior ones. This trend will likely continue as more Japanese researchers spend time abroad and more English-speaking foreign researchers spend time in Japanese research laboratories.

Japan participates in a number of world-wide research activities, such as global warming research, the Space Station Program, research at the European Laboratory for Particle Physics (CERN), and the International Thermonuclear Experimental Reactor Project (ITER).

ITER is an international project supported by Japan, the EU, the US, and the Russian Federation under the auspices of the International Atomic Energy Agency (IAEA). The eventual aim of the project is to build an experimental reactor. Initial conceptual design activity took place at Garching, Germany, between 1988 and 1990. The next phase, the engineering design activity began in 1993 and is expected to last six years. The joint design work is conducted at three sites – Naka in Japan, San Diego in the US, and Garching in Europe. The future of ITER is currently uncertain.

Japan also participates in the International Space Station Program undertaken jointly by the United States, member nations of the European Space Agency (ESA), Canada, Japan and Russia under an Intergovernmental Agreement. Japan's role is to develop the Japanese Experimental Module (JEM), although collaboration has been beset by scaling back of initial designs as a result of the budgetary constraints faced by other partners in the collaboration.

In addition to participating in international research activities initiated by other countries or international organizations, Japan also promotes its own international collaboration programs such as the Human Frontier Science Program (HFSP), and the International Joint Research Program (IJRP). For example, the HFSP was proposed by the Japanese government to support basic research on biological functions. Formally launched at the Venice Economic Summit in 1987, Japan provides the bulk of funding (80 percent), with the remainder contributed by European countries and the US. Three forms of support are available, research grants for projects that involve collaboration between teams in different countries, fellowships for post-doctoral scientists who wish to work in a different country and support for international workshops. It is an interesting reflection of the perceived quality of Japanese research in the area that of eighty-seven fellowships (of which about fifty have been awarded to non-Japanese researchers), only two – both Europeans – chose to pursue their studies in Japan.

IJRP, an international version of the ERATO program (see above), was established with the intention of promoting international collaboration in basic research. An international joint project runs for five years with a total budget of 2–3 billion Yen and usually involves about twenty researchers drawn from industry, academia, and government. Costs are shared on a 50–50 basis, while laboratories are located where they can best serve the needs of the project. Research results are actively disseminated through the open scientific literature and intellectual property rights are shared by the Japan Science and Technology Corporation (JST) and the counterpart organization. There are currently five projects being supported: two involve groups in the US; the others involve groups in France, Germany and Sweden. A project involving researchers from Cambridge University and the University of London was completed in 1995.

As one final example of an international collaboration program originated by Japan, the New Energy and Industrial Technology Development Organization (NEDO) Research Grant to International Joint Research Teams (the NEDO Grant) was established in 1995. The NEDO grant is intended to support international joint research teams which conduct original and innovative basic research in the materials, energy global environment areas. This program provides grants to selected joint research terms to cover research expenses.

The internationalization of Japanese science and technology is a priority of the new Basic Plan. The Plan builds on earlier developments and calls for the promotion of joint R&D with a particular emphasis on global issues and for cooperation with developing countries. The Plan also calls for expanding international exchange programs by, for example, increasing the number of fellowships and improving researcher support systems. Again, if these developments can be sustained, then Japan cannot fail to contribute to, as well as benefit from, a greater involvement in the world science and technology system.

Conclusion

This chapter has examined Japanese efforts to increase the internationalization of its science. Increasing internationalization requires that the quality of Japanese science continues to improve as well as specific measures be introduced to promote internationalization. The new Science and Technology Basic Plan promises to make Japanese science 'more flexible, competitive and open.' Earlier 'experiments' in the way science is managed in Japan provide pointers to the way this might be achieved.

In addition to improving the quality of research undertaken in Japan, such developments should improve the attractiveness of the Japanese research system to foreign researchers. In addition, internationalization is reiterated as a priority under the Basic Plan. The hope is that this process of internationalization will accelerate in the future as a result of increased resources (the Ministry of Finance notwithstanding), increased research quality, and increasing experience with successful international collaborations.

REFERENCES

Arima, Akito (1996) The Evolution of Japan's Research System and its Links with South East Asia. Paper presented at the EU-RESTPOR conference: *Global Comparisons of Regional RTD & Innovation Strategies for Development and Cohesion*, 19–21 September.

Barker, Brendan (1996) *Japan: A Science Profile*. London: British Council.

Basic Law (1995) *Kagaku Gijutsu Kihon Hō* (The Science and Technology Basic Law), Hōritsu Dai-130-go (Law no. 130), 15 November 1995.

Dawkins, William (1996) *Financial Times*, 3 July, pp. 5 and 12.

Geller, Robert (1994) Presentation to the conference: *Nurturing Creativity in Research: Support for Basic Research*, organized by Nature, November 1994, Tokyo.

Gakujutsu Shingikai (ed.) (1992) *21-seiki o tenbōshita gakujutsu kenkyū no sōgōteki suishin hōsaku ni tsuite* (Strategies for comprehensive promotion of scientific research with a view to the twenty-first century). Tokyo: Gakujutsu Shingikai.

Georghiou, Luke (1993) *Evaluating US and Japanese Technology and Cooperation with the EC*. Report to STOA (Science Technology Options Assessment), European Parliament.

Hayashida, Hideki (1996) Science Policy in Japan. In *Science*, vol. 272, p.1567.

Kagaku Gijutsucho (1995) *Kagaku gijutsu hakusho 1995* (White Paper on Science and Technology 1995). Tokyo: Okurasho Insatsukyoku.

Kagaku Gijutsu Kaigi (1996) *Kagaku Gijutsu Kihon Keikaku* (Basic Plan for Science and Technology). Tokyo: Kagaku Gijutsu Kaigi.

Koizumi, Kenkichiro (1993) The Challenge of Joint Research in Japan. In *Science*, vol. 258, p. 7.

NISTEP – Kagaku Gijutsu Seisaku Kenkyusho (1995) *Kagaku gijutsu shihyō. Nihon no kagaku gijutsu katsudō no taikeiteki bunseki* (Science and technology indicators. A systematic analysis of Japanese activities in science and technology). Tokyo: Kagaku Gijutsu Seisaku Kenkyusho (= NISTEP Report no. 37).

Normile, Dennis (1996) New Faculty Grants Program Expands Role of STA. In *Science*, vol. 272, p. 645.

NSF – National Science Foundation (1996) *Science and Engineering Indicators 1996*. Washington, DC: National Science Foundation.

NSF Tokyo Office (1995) *Report Memorandum Number 95–2*. Tokyo: National Science Foundation Tokyo Office.

—— (1996) *Report Memorandum Number 96–6*. Tokyo: National Science Foundation Tokyo Office.

—— (1997) *Report Memorandum Number 97–9*. Tokyo: National Science Foundation Tokyo Office.

Odagiri, Hiroyuki and Akira Goto (1997) *Technology and Industrial Development in Japan*. New York and Oxford: Oxford University Press.

Samejima, Muneaki (1995) *Japan's Newest Basic Law: Support for Basic Research and Technology, Nurturing Creativity in Research*, Conference sponsored by Nature and The Institute of Advanced Studies, Australian National University, Canberra, Australia, 26–28 November 1995.

Sigurdson, Jon (1995) *Science and Technology in Japan*. London: Cartermill.

Swinbanks, David (1993) Japanese Universities are Slow to Welcome Foreign Faculty. In *Nature*, vol. 363, p. 290.

6

ORGANIZATIONAL INNOVATION IN JAPANESE BASIC RESEARCH

From bureaucracy to dynamic network

Ken Kusunoki

Introduction: problems and challenges in Japanese basic research organizations

Recently, sources of competitive advantage of Japan have been shifting from manufacturing to R&D (Porter 1990). Japan is now becoming a leader not only in manufacturing but also in engineering such as product and process development (Okimoto 1986). On the other hand, still lagging behind Western countries, Japanese contributions in the field of basic research are less significant (Sakakibara 1991). One of the most critical problems of Japanese basic research lies in the management styles of basic research organizations in Japan. Most of Japanese basic research organizations, especially universities and national laboratories, can be characterized as bureaucratic organizations which primarily pursue stability as well as continuity of research staffs and activities at the cost of organizational flexibility and fluidity. For example, considering human resource management in Japanese basic research organizations, almost all of them basically apply lifetime employment and uni-dimensional promotion systems based on seniority. Under such bureaucratic management systems, once entering a particular organization, most Japanese researchers stay in the same organizations throughout their careers. Bureaucratic organizations which are now dominant in Japanese basic research have some advantages in the sense that they carry out basic research activities in accumulative, systematic, and continuous ways, in order to learn from and catch up with basic research in Western countries. At the same time, however, bureaucratic organizations obviously pose some serious problems since they limit organizational flexibility and fluidity which are essential to basic research activities.

Considering theoretical aspects of R&D management, much of the existing literature has consistently argued that such a lack of flexibility and fluidity in bureaucratic organizations has negative effects on research performance as a result of the decrease in professional communication among researchers. In R&D activities, it is quite important for professionals to collect and process information from a variety of sources. R&D professionals spend a large amount of their time collecting and processing information through various media such as technical reports, publications, discussion with others and so forth (Allen 1977). Among them, face-to-face oral communication through personal contact is especially important because it permits timely information exchange, rapid feedback, and critical evaluation, as well as the opportunity for real-time recording and synthesis of information (Allen 1977; Tushman 1979). Professional communication can be categorized into two types: internal communication among professionals in the same organization, and external communication with those who belong to other organizations (Katz and Tushman 1979). The importance or effectiveness of either type of communication varies depending on characteristics of R&D tasks. In the field of basic research, external professional communication becomes critically important in terms of research performance, while engineering activities such as development and technical service need internal communication rather than external (Hagstrom 1965; Allen, Tushman, and Lee 1979). One explanation for this heavy reliance of basic research on external communication is that, compared with more organizationally specific tasks such as engineering, basic research can be defined as a more universal task, and therefore it becomes impossible and inefficient to secure information self-sufficiently within one individual organization (Thompson 1967; Mansfield and Wagner 1975; Allen 1977). In this sense, external communication among professionals has been recognized as a key variable in exploring the relationship between organizational behavior and performance in the field of basic research.

However, existing literature shows that as organizations lose flexibility and fluidity, both internal and external professional communication decrease, and finally this tendency results in a significant decline of performance (Pelz and Andrews 1976; Smith 1979; Katz 1982). In the case of basic research, the decrease of external communication accompanied by excessive organizational stability is especially problematic since it leads to a serious decline in research performance. This kind of organizational phenomenon is often called a Not Invented Here (NIH) syndrome (Katz and Allen 1982), and suggests the limits of Japanese basic research within bureaucratic organizations which inevitably exclude organizational flexibility and fluidity. Recently, this kind of organizational problem has been frequently discussed in Japan. Numerous researchers and managers in Japanese basic research organizations have claimed that the bureaucratic management systems foster an introverted mentality among researchers which discourages professional interactions over boundaries of organizations, and becomes an obstacle in strengthening Japanese activities in basic research (Iinuma, Iinuma, and Kusunoki 1991).

On the other hand, however, some interesting organizational experiments to overcome such Japanese problems in bureaucratic basic research organizations have come into operation. Recently, basic research organizations with a new management style have been established in Japan. These newly created basic research organizations are characterized by exceptional flexibility and fluidity in their management system. I have labeled them 'dynamic network organizations.' The creation of dynamic network organizations has an aspect of 'organizational innovation' in basic research in Japan. The most substantial case of the dynamic network organization is ERATO (Exploratory Research for Advanced Technology), which was founded by the Science and Technology Agency (STA) of the Japanese government.

Paying attention to the organizational innovation in Japanese basic research, this paper explores the effectiveness of emerging dynamic network organizations and problems in bureaucratic organizations in Japanese basic research in terms of professional communication and performance. The research is based on a comparative case analysis of ERATO and one of the most distinguished basic research institutes in Japan, a laboratory of MITI which is a typical representative of bureaucratic organizations (see Table 6.1).

Research method

Comparative case analysis

The method used in this research project is a comparative case analysis based on 'theoretical sampling.' In other words, this research conducts an in-depth analysis of a comparable, but sharply contrastive pair of basic research organizations which were carefully selected in the light of theoretical relevancy. The purpose of this research is to inductively derive a series of hypotheses out of empirical data, rather than to rigidly test specific hypotheses. Since this research is orientated toward hypothesis-finding, a comparative case analysis based on theoretical sampling is more effective as a research method than others such as survey research investigations with large samples obtained from random sampling techniques (Glaser and Strauss 1967). This method enables a researcher to shed light on critical differences between the cases, and to inductively explore various factors underlying the differences (Eisenhardt 1989; Kanai 1989). Furthermore, since this kind of organizational innovation in Japanese basic research, the creation of the dynamic network organization, is precisely a quite contemporary and experimental phenomenon, it is safe to say that this research focuses on a revelant and unique case. In such cases, a case analysis can be a particularly effective research method (Yin 1984).

The comparative case analysis was carried out by setting a conceptual taxonomy which contrasts the two ideal types of research organizations, the bureaucratic organization and the dynamic network organization, from the perspective of managerial characteristics in terms of organizational flexibility and fluidity. As shown in the next section (Table 6.2), the taxonomy consists of six dimensions: (1) recruitment and staffing, (2) assignment, (3) employment, (4) promotion, (5) inter-divisional transfer, and (6) sustainability of research activities. This taxonomy itself can be considered as a generic taxonomy including theoretical constructs to generally contrast various research organizations from the perspective of organizational flexibility and fluidity resulting from management systems. However, the taxonomy can be positioned at an empirical level as well as at a theoretical level, since it was empirically derived from the actual observation and comparison of the two organizations selected for the research.

Research setting

The organizations selected for the comparative case analysis are ERATO as an example for the dynamic network organization, a basic research organization sponsored by STA, and a MITI (Ministry of International Trade and Industry) Laboratory (here also referred to as MITI-L, a provisional name) as a representative of the bureaucratic organization.

In the first stage of the study, pilot research including interviews and field-observation was conducted in order to select an appropriate pair of cases in the light of theoretical relevancy. In the pilot research, I collected information concerning ERATO, International Frontier Research, Center for Photo-Dynamics Research, Kanagawa Academy of Science and Technology, and Osaka Bioscience Institute. These basic research organizations could be recognized at the time of research to have substantially introduced flexible and fluid management systems such as time-limited research groups, employment systems based on individual-based contracts, and so forth. Among them, I selected ERATO as a representative of the dynamic network organization because it was the pioneering and the most substantial case among the dynamic network organizations. According to interviews with managers of the four remainders, they could be regarded as followers of ERATO in the sense that they had been consciously learning from ERATO's experience.

At the same time, I explored an appropriate case of the bureaucratic organization which could be considered as a theoretically comparable counterpart. As a result, MITI-L was selected as a representative of a bureaucratic organization. The reasons can be defined as follows. First, as discussed in the next section, MITI-L stands in striking contrast to ERATO in terms of organizational flexibility and fluidity. Along the six dimensions included in the taxonomy, MITI-L appears to hold typically bureaucratic management systems.

Second, apart from such organizational and managerial characteristics,

MITI-L is so similar to ERATO that the comparative analysis can avoid effects from factors that were not directly related to the interest of this research. The task characteristic of MITI-L is consistent with that of ERATO in the sense that both were conducting basic research, and the specialties of research activities within MITI-L are similar to those of ERATO. Both organizations are divided into several research departments including twenty to thirty researchers. In both organizations, research groups (the smallest organizational units of research activity) have specific research themes and consist of five to ten researchers. Furthermore, there is no substantial difference in the amount of research funds. The average budget per researcher is approximately $150,000 in both organizations. In sum, between ERATO and MITI-L, there is little difference in the aspects mentioned above, while they offer remarkable contrast in managerial characteristics on which the comparative analysis focused.

Finally, MITI-L is precisely one of the most distinguished basic research organizations in Japan, and it has established a remarkable reputation in the field of basic research. In order to evaluate the effectiveness of ERATO's organizational experiment in comparison with a bureaucratic organization, it would be important to set a strong counterpart.

Data

The data used in the comparative case analysis was collected from three sources. Accessing multiple sources of data, this research pursues the benefits of methodological triangulation (Fielding and Fielding 1986). First, qualitative data was collected through intensive, semi-structured, open-ended interviews with researchers and managers in both organizations (n = 44). The interviews typically lasted between one and two hours, but some lasted as long as four hours. In some cases, even after completing the first interviews, further information was collected by additional interviews.

Second, archival data such as published or unpublished internal reports on ERATO and MITI-L was also collected. Importantly, both organizations had conducted general surveys about their research activities and researchers. Archival data also included the pamphlets and brochures prepared by both organizations.

The third source of data is a questionnaire survey. Interviewing managers and researchers from each organization, the contents and wordings of the questionnaire were checked and elaborated in advance in order to fit respondents' native terms as much as possible. Questionnaires were sent to researchers who had spent more than one year in either ERATO or MITI. In most cases, questionnaires were sent by asking a few persons in each organization to distribute them randomly and collect them through mail boxes. Some questionnaires were handed out and collected directly during the interviews. The data obtained from these questionnaires is used mainly for exploring the differences between ERATO and MITI-L in the frequency of professional communication as well as researchers' performance.

The number of respondents of the survey into ERATO and MITI-L was 108 and 119, respectively. The average age of respondents is 33.8 (standard deviation = 4.63) in ERATO and 37.0 (6.54) in MITI-L. Although respondents in ERATO are three years younger than those in MITI-L on the average ($t = -4.29$; $p < 0.01$), as shown in the next section, it can be assumed that this difference in age reflects differences in management systems concerning recruitment and staffing. Looking at the distribution of respondents' organizational positions, 11.1 percent ($n = 12$) of ERATO's respondents are research leaders, 21.3 percent ($n = 23$) are research sub-leaders, and 67.6 percent ($n=73$) are common researchers. In MITI-L, 14.3 percent ($n=17$) of respondents are research leaders, 22.7 percent ($n = 27$) are sub-leaders, and 63.0 percent ($n = 75$) are common researchers. There is no statistically significant difference between the two organizations (chi-square = 0.67; $p = 0.71$).

In the comparative case analysis, these three types of data were used to complement each other. Qualitative data obtained from interviews and archives was typically used in generally grasping and describing the difference between ERATO and MITI-L in organizational and managerial characteristics. Quantitative data from the questionnaire survey was used for statistical analyses which was to explore the degree of the differences and relationships of variables. Again, qualitative data was heavily used in interpreting the results of the statistical analyses.

Bureaucratic organization versus dynamic network organization: overall description of managerial characteristics in ERATO and MITI-L

Setting the taxonomy which contrasts the bureaucratic and the dynamic network organization, this section describes the overall comparison of organizational characteristics and management systems between ERATO and MITI-L. Based on the taxonomy which focuses on the flexibility and fluidity in management systems, Table 6.1 shows the summary of the comparative description of the management systems in ERATO and MITI-L. It is important to mention again that this taxonomy is by nature a conceptual framework contrasting the bureaucracy and the dynamic network. The cases of ERATO and MITI-L are considered as empirical references to the two ideal types of research organization. Framed in this way, the bureaucratic and the dynamic network organization are extreme points along the dimension of flexibility and fluidity in management systems, and therefore, the comparison of ERATO and MITI-L consistently offers striking contrasts. In fact, there is a spectrum of research organizations between both types. It would be

Table 6.1 Comparison of managerial characteristics at ERATO and MITI-L: stable hierarchy and dynamic network

Dimensions	*MITI Laboratory: stable bureaucracy*	*ERATO: dynamic network*
Recruitment/ staffing	intensively recruiting researchers once a year (in April)	always open to applicants
	recruiting directly from schools	open to every applicant (including doctoral students and foreigners)
	criteria for assessment and screening are quantitative scores of formal examinations	assessment and screening by intensive interviews with a project leader
Assignment	according to researchers' academic backgrounds and disciplines	according to researchers' proposals for their future research in a project
Employment	lifetime employment	full-time employment contract (for five years at the longest)
Promotion	seniority-based unidimensional	capability-based project leaders are recruited from outside
Intra-organizational transfer of researchers	none (no system/rules)	based on researchers' requests (usually, at the time of renovating employment contracts)
Longevity of research group	open-ended (no system/rules limiting research terms)	projects are compulsorily finished and dispersed after five years
Key words	stability/continuity	flexibility/fluidity

Source: Author's composition.

safe to say that most research organizations would fall somewhere between ERATO and MITI-L.

ERATO: A dynamic network organization

ERATO was founded in 1981 by the Japanese government, STA. As a government-sponsored organization for basic research, it is emphasized that research activities in ERATO are not application-oriented. At the time of this study, ERATO runs sixteen research departments (called 'projects' in ERATO's terminology). The fields of the research projects cover a wide range of natural sciences including physics, chemistry, and biology. Each project is directed by a project leader, and includes three research groups directed by group leaders.

In association with organizational flexibility and fluidity, the most distinctive characteristic of ERATO lies in the point that every project is subject to 'the five-year rule.' All projects are time-limited, and automatically finished and dispersed without exception after five years. One of the managers of ERATO explained the intention behind the five-year rule as follows.

> 'Five-years might seem too short in the light of Japanese common sense in the field of basic research. But the primary purpose of ERATO is to make research activities more flexible and fluid in order to overcome the evils in traditional Japanese basic research organizations which are far too stable. In Japan, we now need some compulsory systems to exclude the naturally emerging orientation of researchers toward over-stable, over-comfortable situations.'

Researchers are extensively recruited through public advertisement. Those who wish to join ERATO are directly and intensively interviewed by a project leader for assessment and screening. In the screening processes, project leaders especially consider applicants' future research plans, rather than their past achievements, career backgrounds, or academic disciplines. Consequently, each project comes to consist of a large variety of researchers in terms of career backgrounds or disciplines. ERATO is open to researchers in national institutes, companies, or universities, including foreigners, postdoctoral researchers, and graduate students. Even in the course of each project work, researchers can be recruited through the same procedures in response to applications. Thus, the timing of recruitment is not specified and, in principle, ERATO's projects are always open to applicants. Although there is no formal condition on eligibility of applicants, relatively young researchers under the age of 35 are encouraged to apply.

Researchers are employed based on time-limited contracts with ERATO. Researchers are typically given two-year contracts by ERATO, but the arrangements can be shorter or longer – up to five years – according to researchers' requests. Every researcher is forced to engage fully in the research in ERATO, and therefore researchers in ERATO can not have concurrent posts in other organizations. In other words, those who are going to join ERATO have to formally resign from a former organization in advance, and they must hunt new outside posts after completing their work at ERATO. However, researchers from companies can be sent to join ERATO as 'delegated researchers' without resigning from their companies, and they can return to their mother companies after finishing their research in ERATO's project. But, even in this case, researchers must engage full-time in ERATO's projects. It is emphasized that the research itself must be independent of interests of other organizations, and that ERATO is by no means an organization for joint research between multiple organizations. Checked carefully in the process of recruitment, researchers from companies are prohibited from bringing into ERATO their work assigned by their mother companies. As described above, management systems in ERATO are designed to enhance individual-oriented research activities as well as organizational flexibility and fluidity.

A particular research theme is assigned to each researcher in accordance with his or her own research proposal. In each project, allocation of research funds is fully delegated to a project leader or group leaders. Research funds or resources can be flexibly allocated to researchers, according to the attractiveness of research proposals by project members. In other words, it is primarily necessary for researchers to submit attractive proposals to their leader, if they are to secure more resources. Researchers' annual salaries are decided not on seniority, but on the project leaders' evaluation about the individual researchers' capability, potential, and contribution to the particular project. Promotion of researchers to group leader or sub-leader is also decided by the project leader on ability.

Because a project leader has extraordinarily strong influence on project work, management staff of ERATO consider the selection of project leaders as one of the most important decisions. Frequently attending academic meetings or conferences, first they list up preferable candidates, and then ask them to prepare research proposals. A project leader is finally decided by the advisory board consisting of twenty outside experts. Among sixteen project leaders, eight joined ERATO from universities, four from national laboratories, and four from companies. A new project leader is always staffed by recruiting from outside at the starting point of the project. In this sense, there is no internal promotion in the course of project work, such as from a group leader to a project leader. Only project leaders can hold concurrent posts in their mother organizations.

In sum, ERATO is a typical dynamic-network-type of organization which pursues extraordinary flexibility and fluidity at the cost of stability and continuity of research activities. It is widely recognized that, compared with Japanese basic research organizations, universities or other basic research organizations in Western countries hold in general more flexible and fluid systems. However, it would be important to mention that ERATO holds much more flexible and fluid systems than most Western organizations. For instance, a group leader who had spent more than ten years at an American university before joining ERATO referred to the uniqueness of ERATO as follows:

> 'Talking about ERATO's systems, for example, management systems such as contract-based employment, or capability-oriented evaluation or rewards, may be close to the so-called American system. But ERATO is nothing but a social or organizational experiment, which

is also different from American universities. An entirely new project starts without any basis by gathering various researchers including those in companies, and after five years, it is forced to be dispersed. And then, another project gets started. In this way, ERATO has unusually dynamic systems obviously including quite experimental ideas. We must be guinea pigs for this social experiment named ERATO.'

MITI-L: A bureaucratic organization

As mentioned earlier, MITI-L is one of the most prestigious basic research organizations in Japan. It is divided into more than twenty research departments, each of which has a particular specialty. Similar to ERATO, a department typically consists of three or four research groups including five to ten researchers, each of which is directed by a group leader and a few sub-leaders.

In MITI-L, in contrast with ERATO, researchers are recruited only once a year, at the beginning of the organizational calendar. Those who want to enter MITI-L are assessed and screened through standardized, formal examinations which are primarily meant to evaluate applicants' knowledge on particular specialties. In the process of screening, only quantitative scores of the examinations are taken into consideration and no qualitative assessments such as direct interviews are conducted. In principle, MITI-L is open only to new graduates who have just received academic degrees from their universities, while ERATO is always open to various applicants. Once applicants are employed, MITI-L guarantees lifetime employment to all employees without exception.

Researchers recruited through this process first go through a short-term introduction training, and then are assigned to particular departments based on their requests. In the process of assignment, however, researchers have no opportunity to specify their requests based on detailed information regarding their actual research activities such as research themes on which they wish to work. Consequently, a newly employed researcher is usually assigned to a particular group in a particular department, according to his or her formal discipline in a graduate school. A research leader in MITI-L explained this kind of assignment as follows:

'Even if we had information about their research interests in detail, it would be difficult to consider that in the actual decision of assignment. This is because each group has been already working on a particular theme for a long time, consistently as a group, without regard to freshmen's proposals. In this sense, it is usually difficult for a group to always take freshmen's various claims into consideration. If a freshman tried to bring his own research theme into a group, the

> research work in the group would face disorder. We recruit a freshman not for doing his or her own research, but for doing our research. I don't think this way of doing things is always good, but under the conditions in our lab, it appears to be the most efficient to assign them by looking at their discipline in their universities.'

In ERATO's case, the decisions concerning the assignment of research themes and allocation of research funds or resources depend primarily on the attractiveness of the individual researcher's proposal to the research leader. In MITI-L, however, such decisions tend to be made by consensus among researchers. As the interview data quoted above also suggests, in MITI-L, research themes and resources are fixed in advance, in accordance with particular organizational goals at the level of departments or groups, not with researchers' proposals on their own initiatives. In this way, research activities in a research group within MITI-L are strongly coordinated and structured by formal hierarchical mechanisms within the organization.

In MITI-L, achievements of individual scientists are evaluated by a research leader of his or her department. But such evaluations are done only in a vague, informal style, and criteria of evaluation are not standardized at all. This may be because the evaluation itself has little meaning in the sense that it is not substantially related to actual aspects of management practices, such as promotion or salary. In MITI-L, seniority and tenure are precisely the most important criteria concerning the decision of promotion and salary. Every researcher can be promoted to a sub-leader in a research group at the age of 33, on the average. However, the greatest deviation in the age of the promotion can not be more than two years. Under this uni-dimensional promotion system, a researcher in MITI-L is promoted from common researcher to research sub-leader and then to research leader, by taking the place of his or her former supervisor. Furthermore, there is no formal system or rule concerning inter-departmental transfer of researchers. A researcher in MITI-L conducts research in only one research group throughout his or her career, in general. Those who have experienced transfers to other departments or groups are exceptions.

In MITI-L, there is neither a formal system or institution for the reformation of research departments, nor a formal regular evaluation system for checking the performance of departments or groups, while ERATO has the five-year rule. Organizational reformation at the departmental level such as the creation of new research departments or integration/abolition of existing departments can be typically realized only when the organization of MITI-L as a whole undergoes revision. In the past, however, this kind of organizational transformation was realized only exceptionally, in 1970 and 1990 with a twenty-year interval.

In sum, in sharp contrast to ERATO, management systems within MITI-L are designed to primarily and consistently pursue stability of research staff and activities. In other words, MITI-L can be regarded as the typical bureaucratic organization which hampers organizational flexibility and fluidity. While the comparative description of managerial characteristics in both organizations mainly depends on qualitative data from interviews and archives, some quantitative data from the questionnaire survey also show the differences in organizational characteristics between ERATO and MITI-L. As shown in Table 6.2, researchers' perception of organizational characteristics also indicates significant differences between both basic research organizations along these dimensions, which are consistent with the above description. Researchers in ERATO are more competitive ($t = 3.92$; $p < 0.01$) and individual-oriented ($t = 5.11$; $p < 0.01$) in doing their work, and at the same time, they are evaluated more clearly ($t = 6.14$; $p < 0.01$) based on more capability-oriented criteria ($t = 3.82$; $p < 0.01$). On the contrary, research activities in MITI-L are directed and coordinated by more group-oriented decision-making ($t = -6.06$; $p < 0.01$). The variety of researchers' backgrounds or orientations in ERATO is greater than that in MITI-L ($t = 3.62$; $p < 0.01$). Looking at the distribution of universities from which respondents graduated, the concentration ratio (leading five universities) is greater in MITI-L (75.6 percent) than in ERATO (48.1 percent). More than 75 percent of respondents (n=119) of MITI-L graduated from five Japanese top schools in the field of the natural sciences, University of Tokyo (n = 40), Tokyo Institute of Technology (n = 20), Kyoto University (n = 12), University of

Table 6.2 Comparison of organizational characteristics at ERATO and MITI-L

In your organization . . .	*ERATO* (*n* = *108*)	*MITI-L* (*n* = *119*)	*t-value*
Researchers are very competitive	4.81 (1.16)	4.19 (1.22)	3.92***
Research work is strongly individual-oriented	5.15 (1.09)	4.34 (1.28)	5.11***
Decision-making is strongly group-oriented	3.28 (1.32)	4.42 (1.52)	−6.06***
Assessment is strongly ability-oriented	5.01 (1.08)	4.35 (1.46)	3.82***
Assessment is clearly fed back to a researcher	4.75 (1.09)	3.75 (1.32)	6.14***
There is a great variety of researchers' backgrounds and/or orientations	5.53 (1.09)	4.89 (1.15)	3.62***

Source: Author's composition.

Notes:

a Each item is measured by a Likert-type seven-point scale (one for strongly disagree, seven for strongly agree).

b Numbers in parentheses are standard deviations.

c Significance is indicated by asterisks: ***$p < 0.01$.

Osaka (n = 12), and Tohoku University (n = 10), all of which are national universities.

The survey also covered a question regarding the task characteristics or the contents of research within both organizations. According to the result of a t-test for mean difference, respondents' perceptions indicate that both ERATO and MITI-L are 'basic-research-oriented' organizations to the same degree. I measured 'task orientation toward basic research' by asking a question with a seven point scale that 'in your organization, short-term applicability of research themes hardly comes into question.' The results are 4.85 (s.d. = 1.57) in ERATO and 4.46 (s.d. = 1.54) in MITI-L. A two-tailed t-test shows that there is no significant mean difference between the two organizations in terms of research content, which would ensure comparability between ERATO and MITI-L in the areas of research content, time orientation, and 'advancedness.'

Professional communication and performance

The striking contrast regarding organizational and managerial characteristics between ERATO and MITI-L suggests that there are also some critical differences between both organizations in professional communication among researchers. As this paper reviewed in the beginning, existing literature of R&D management has consistently suggested that the lack of organizational flexibility and fluidity in bureaucratic organizations has significant negative effects on performance through the decrease of external professional communication which is essential for basic research activities. Thus, MITI-L as the representative of bureaucratic organizations can be assumed to face a series of problems like the NIH syndrome. On the contrary, the organizational innovation of ERATO as the dynamic network organization can be assumed to overcome such problems by activating professional communication with researchers outside of the organization which would result in higher performance.

Differences in professional communication and performance

In analyzing the differences in frequency of researchers' professional communication, I used nine variables included in the questionnaire which were assumed to relate to professional communication:

1 frequency of professional communication with researchers within the organization who belong to the same research group;
2 frequency of professional communication with researchers who supervise the respondent;
3 frequency of professional communication with researchers within the organization who belong to different research groups;

4 frequency of professional communication with researchers outside of the organization with the same research specialty;
5 frequency of professional communication with researchers outside of the organization with different research specialties;
6 number of times of attendance at academic meetings or conferences during the past one year;
7 number of times of attendance at formal meetings within the group, the department, or the organization during the past one year;
8 average number of times of lunch with colleagues in a week;
9 engagement in joint research with researchers outside of the organization.

As for the communication variables 1–5, each variable was measured by a Likert-type seven-point scale (one for 'less than once in a year', four for 'once in a week', and seven for 'many times in a day'). Variables 6–8 were measured by the actual number of times. The variable 9 was a dummy variable to be measured by one for 'Yes' or zero for 'No.' As a result of factor analysis, there emerged five factors with eigenvalues more than 1.00 (see Appendix Table 6.7). Checking factor loading after Varimax rotation, I used the score of Factor 1 as a measure called 'external communication', and the score of Factor 2 as 'internal communication' in the following analysis.

Table 6.3 shows the results of a T-test for the differences in mean frequencies of internal and external communication between ERATO and MITI-L. As expected, there is a significant difference in external communication with the external communication measure of ERATO being greater than that of MITI-L. Although the difference is not significant ($p = 0.14$), the internal communication measure of ERATO is also greater than that of MITI-L. These findings show that, compared to MITI-L as the typical bureaucratic organization, the dynamic network organization of ERATO includes a tendency to facilitate researchers' communication in general, and such a ten-

Table 6.3 Comparison of mean communication frequencies at ERATO and MITI-L

Communication measures	*ERATO (n = 108)*	*MITI-L (n = 119)*	*t-value*
Internal communication	0.10 (0.89)	−0.09 (1.08)	1.47
External communication	0.12 (1.11)	−0.11 (0.86)	1.71*

Source: Author's composition.
Notes:
a Factor scores are used as communication measures.
b Numbers in parentheses are standard deviations.
c Significance is indicated by asterisks: *p < 0.10.

dency is significant in the case of external professional communication. On the contrary, external communication appears to be suppressed in the bureaucratic organization.

Both existing organizational theory and the interview data suggest some complementary explanations for this difference in external communication. First by creating extraordinarily fluid work situations, the dynamic network organization can be assumed to escape from phenomena such as behavioral stability or selective exposure to outside information, which negatively relate to external communication. Researchers interacting over a long period under stable work situations will develop standard work patterns that are familiar and comfortable, and come to rely more and more on their customary ways of doing things (Thompson 1967; Katz 1978; Pfeffer 1983). It is posited that one of the consequences of such behavioral stability is that researchers become less receptive toward communication that may not fit their stabilized patterns of behavior, and become increasingly isolated from outside sources of relevant information, due to the lack of organizational fluidity (Katz 1980). In the case of ERATO where projects are started and compulsorily finished according to the five-year rule and management systems such as work assignment and staffing are practiced in flexible manners, there might be no room for such phenomena to occur at all.

Second, the dynamic network organization can be assumed to make external communication easier, since it consequently comes to consist of a great variety of researchers in terms of career backgrounds and orientations enough to enhance their skills to communicate with researchers outside of the organization. In the dynamic network organization, there necessarily emerges a large variety of researchers, because they have been collected into a research group through flexible and fluid recruiting/staffing systems. Given this situation, each researcher is forced to establish certain skills to communicate efficiently with colleagues who have a variety of different backgrounds and orientations. If they have consciously or unconsciously established skills to communicate with those who have different backgrounds and orientations in the context of daily work, such skills may also contribute to exclude mental and/or verbal impediments in external communication, and may result in facilitating external communication. According to the interview data, many researchers in ERATO consistently pointed out that 'the universal linguistic system' was one of the important characteristics of ERATO. A researcher typically noted as follows:

> 'This group consists of various people with different career backgrounds, different disciplines, or different ways of thinking. So, everyone comes in need of learning certain techniques to communicate with different people in terms of background or orientation, especially in the early stage of project work. I feel that people in ERATO, including myself, have gradually come to speak or express

> their ideas in a more universal way even in daily situations, in order to overcome such gaps in communication. Now I think I was using more local terms in discussing with colleagues before joining ERATO, just because I could communicate smoothly and quickly by using more local terms such as jargons or so. There was no problem like that here at ERATO, but at the same time, that would be some kind of barrier of interactions beyond particular labs. I think that this is the critical difference between ERATO and other labs.'

Third, it is assumed that even after joining the dynamic network organization, it can be easier for researchers to maintain close connections with people working for organizations to which they once belonged, thus they can secure many more opportunities to communicate with the outside through these connections. In ERATO's case, researchers can stay at ERATO for five years at the most, because each project is subject to the five-year rule. According to the interviews, many researchers in ERATO mentioned that they could work on the project of ERATO without losing close ties with their ex-colleagues, because it had not been such a long time since they had left the former organizations. Moreover, researchers with various career backgrounds bring their own networks they established before joining ERATO. In consequence, making the most of others' various professional networks as well as their own, researchers may be able to greatly increase the number of opportunities or channels to communicate with outside professionals. Given organizational flexibility and fluidity in terms of project longevity and staffing, the dynamic network organization may enable researchers to easily access a large variety of outside information sources by increasing communication channels through various professional networks which they bring into the organization.

On the other hand, it can be assumed that researchers in the bureaucratic organization may inevitably come to suffer from the behavioral stability and the insufficient skills for external communication, as well as relatively limited opportunities and channels for external communication. These factors suppressing external communication may be underlying the finding that the frequency of external communication is less in MITI-L than in ERATO.

Given the existing studies on professional communication and performance, the difference in external communication shown above also suggests the significant difference in researchers' performance between ERATO and MITI-L, since external communication is a critically important factor for researchers to achieve high performance. Performance of researchers in both organizations was measured by counting the number of respondents' publications. Since they are engaged in basic research, the number of publications is one of the most important performance indices.

The questionnaire includes four variables relating to the number of publications:

1 publications on domestic (Japanese) academic journals;
2 publications on international academic journals;
3 papers presented at domestic academic meetings or conferences; and
4 papers presented at international academic meetings or conferences.

These four variables indicated by each respondent were aggregated. In aggregating these variables, each paper or article included in either 1 or 3 was counted as 0.5 point, while each paper in 2 or 4 was counted as 1 point. This was because a number of researchers interviewed noted that there was some difference in the significance of papers between international and domestic journals or meetings. Typically, they claimed that a single paper in an international journal or meeting would by and large be equal to two papers in domestic journals or meetings. In the analysis, I used the number of publications aggregated in this way as a performance measure.

As shown in Table 6.4, there is a statistically significant difference in the researchers' performance between ERATO and MITI-L in the predicted direction. On the average, ERATO's performance (3.99) is significantly higher than MITI-L's (2.77). The difference in researchers' performance appears consistent with the result of the comparative analysis on professional communication. These findings suggest that the difference in the frequency of external communication between ERATO and MITI-L would actually relate to the difference in researchers' performance.

In addition, I also surveyed the other researchers' performance in the one year before joining ERATO by using the same performance measure, in order to compare that with the performance of ERATO's original respondents ($n = 118$) shown in Table 6.4. This additional survey was carried out separately from the original questionnaire survey, and the sample was limited to those who had been working for the four projects in ERATO lasting for less than one year ($n = 24$). As a result, the mean performance of twenty-four respondents in the one year before joining ERATO turned out to be 2.38 (standard deviation = 2.09) which is significantly lower than the performance of ERATO's respondents in the original survey ($t = -3.77$; $p < 0.01$). Interestingly, it is also lower than that of respondents in MITI-L ($t = 1.96$; $p < 0.10$). These findings suggest that those who joined ERATO may not

Table 6.4 Comparison of mean performance at ERATO and MITI-L

	ERATO (n = 108)	*MITI-L (n = 119)*	*t-value*
Mean performance	3.99 (3.49)	2.77 (3.35)	2.67***

Source: Author's composition.
Notes:
a Standard deviations are given in parentheses.
b Significance is indicated by asterisks: ***$p < 0.01$.

have been so excellent before, and ERATO's relatively high performance may be brought about not simply by staffing with superior researchers, but by realizing their potential and enhancing their capabilities.

Combining the findings so far, it could be assumed that ERATO as the dynamic network organization actually facilitates researchers' external communication by increasing organizational flexibility and fluidity, and the facilitated external communication would result in relatively high performance. At the same time, the excessive organizational and managerial stability in MITI-L could be assumed to hinder external communication which would be critically important for basic research, and in consequence, result in lower performance.

Relationship between professional communication and performance

In order to analyze the above assumption regarding the differences in both professional communication and researchers' performance between ERATO and MITI-L, I examined the impact of professional communication on the performance in each organization in more detail. Table 6.5 shows the results of regression analysis for ERATO and MITI-L separately. In the regression model, the dependent variable is the researchers' performance and the independent variables are the internal and external communication. Interestingly, the results obtained from regression analysis show that communication measures do not relate equally to performance of ERATO and MITI-L. As predicted, external communication has a significant positive impact on researchers' performance in ERATO, while internal communication has no

Table 6.5 Multiple regressions on performance by internal and external communication at ERATO and MITI-L

Independent variables	*ERATO* (*n* = *108*)	*MITI-L* (*n* = *119*)
Internal communication	0.51 (0.33)	0.72*** (0.26)
External communication	1.69*** (0.26)	−1.48*** (0.33)
Constant	3.73	2.67
R-square	0.36	0.27
F-value	30.40***	22.11***

Source: Author's composition.

Notes:

a Factor scores are used as communication measures.

b Numbers are regression coefficients. Numbers in parentheses are standard errors.

c Significance is indicated by asterisks: ***$p < 0.01$.

significant effect on performance. These findings are consistent with the arguments so far.

According to the interview data, most researchers in MITI-L, especially young researchers below the age of 35, clearly pointed out that more or less isolated from outside professional information sources, they were in need of more external communication and one of the most serious problems in MITI-L would lie in the insufficiency of external communication. Surprisingly, however, external communication has a significantly negative impact on researchers' performance in MITI-L, while internal communication holds a significantly positive impact. Since the findings show that external communication by researchers in MITI-L hinders performance contrary to the proposition, now it becomes impossible to simply assume that the relatively low performance of MITI-L may result from less frequent external communication.

The most simple explanation for this negative association between external communication and performance is that researchers in MITI-L would somehow communicate with outside researchers only to access meaningless or inappropriate information. In the light of existing studies, however, this kind of explanation must be far from reality. In fact, the interviews with researchers in MITI-L confirmed that nearly all of them consider external professional communication one of the most useful means in generating ideas or solving problems. If so, the alternative hypothesis may be that, in the case of MITI-L, such a negative association would result from some organizational mechanisms intervening in the relationship between professional communication and performance. This hypothesis assumes that the mysterious negative relationship between external communication and performance exists only due to the existence of intervening organizational mechanisms in MITI-L, although external communication would be still important for researchers in MITI-L. In examining this hypothesis, it may be helpful to raise a question: 'Who communicates more frequently than others?' I analyzed differences in frequency of internal and external communication and performance across organizational positions of researchers, separately for both organizations. Table 6.6 shows the results of variance analysis using one-way ANOVA to test for significant mean differences across three position categories.

As for ERATO, the two communication measures reveal no significant differences across the three position categories. The results indicate that researchers in ERATO equally communicate with professionals inside and outside the organization, regardless of their positions. As for MITI-L, on the other hand, both communication measures reveal sharply significant differences across the three positions. Table 6.6 shows that common researchers communicate more frequently with colleagues within MITI-L than leaders and sub-leaders. On the contrary, researchers at higher positions, especially research leaders, communicate much more frequently with outside

Table 6.6 Mean communication and performance as a function of organizational position at ERATO and MITI-L

Organization	*Measures*	*Categories of organizational position*			
		Common researcher	*Sub-leader*	*Leader*	*F-value*
ERATO (n = 108)		n = 73	n = 23	n = 12	
	Internal communication	0.03	0.22	0.26	0.59
	External communication	0.01	0.17	0.64	1.69
	Performance	3.28	5.19	5.95	5.10***
MITI-L (n = 119)		n = 75	n = 27	n = 17	
	Internal communication	0.13	−0.43	−0.53	4.66**
	External communication	−0.35	−0.001	0.80	15.67***
	Performance	3.52	2.14	0.47	6.99***

Source: Author's composition.
Notes:
a A one-way ANOVA is used.
b Factor scores are used as communication measures.
c Significance is indicated by asterisks: **$p < 0.05$; ***$p < 0.01$.

professionals, and in consequence, external communication is strongly biased to research leaders. In short, while organizational positions have little to do with professional communication in ERATO, in the case of MITI-L there is a certain kind of 'division of labor' between common researchers, sub-leaders, and leaders, in terms of professional communication.

Furthermore, performance measures of both ERATO and MITI-L indicate strongly significant differences across the three position categories, as shown in Table 6.6. Interestingly however, the direction of the differences in ERATO is sharply opposite to that in MITI-L. In MITI-L, the performance of common researchers is much higher than that of other position categories, and research leaders indicate the lowest performance. In sharp contrast, the performance of leaders or sub-leaders in ERATO is much higher than that of their subordinates. These results suggest that there are critical differences in the role of a research leader between ERATO and MITI-L. A research leader in ERATO is as much a star researcher as a formal director of his research project or group. In the case of ERATO, as already suggested in the former section of this paper, the base of leaders' power or leadership may lie in their own research capabilities. In contrast, a research leader in MITI-L is nothing but a managing director of the research group who is hardly engaged in actual research activities. His leadership may come from the formal authority in the organizational structure, not from his own capability to participate in current research activities.

The findings so far suggest that external communication in MITI-L would be realized by a limited number of research leaders who function as the windows of research groups for information from outside the organization. It could be further assumed that such research leaders would communicate with outside people not because they themselves need outside information in doing actual research activities, but simply because they are formal representatives of their research groups. On the contrary, researchers at different positions in ERATO equally contact with outside information sources in order to satisfy their own needs for getting new ideas and information which is relevant to their actual project work.

While each researcher in ERATO communicates directly with outside professionals, it is assumed that there could be a two-step information flow in MITI-L in terms of external communication, in the sense that a research leader in MITI-L would play the role of a gatekeeper (Allen and Cohen 1969). This assumption is consistent with the findings that there is a significant difference in performance measure between ERATO and MITI-L, because the existing literature has confirmed that external communication via gatekeepers negatively affects performance in the field of basic research (Allen, Tushman, and Lee 1979; Tushman and Katz 1980). Interestingly however, the data obtained from the survey reveals that there is a significant negative correlation between external and internal communication in MITI-L (correlation coefficient = – 0.40; $p < 0.01$), while there is a strong positive correlation in ERATO (correlation coefficient = 0.39; $p < 0.01$). These results show that internal and external communication are split in MITI-L, while researchers in ERATO communicating frequently with outside professionals also communicate with colleagues at the same time. In MITI-L therefore, it can be assumed that outside information collected mainly by research leaders would not circulate through their groups, and surprisingly, they do not perform even as gatekeepers at all. This split between internal and external communication in MITI-L may relate to the argument that research leaders in MITI-L are managing directors rather than researchers actively working on actual research activities. In any case, these findings suggest that outside information mainly collected by leaders would not actually circulate within research groups in MITI-L, and eventually, such outside information would be just collated by research leaders in vain. On the contrary in ERATO, outside professional information is more actively internalized regardless of organizational positions, and at the same time, such internalized professional information is actively circulating through groups to be used for idea generation or problem-solving in actual research activities.

Furthermore, there would be good reason for the fact that external communication by common researchers in MITI-L is much more limited than that by leaders or sub-leaders. Given the organizational structure where research leaders are in charge of a significant part of external communication as formal representatives of research groups, some qualitative data from interviews also suggest that it would be wiser for common researchers not to communicate so frequently with outside professionals, since such behavior would be threatening and disturb the organizational order of authority. In the interviews, some young researchers in MITI-L mentioned that, feeling 'some kind of invisible barriers,' they had naturally kept away from external communication, although they were in need of a variety of outside information at the same time. Some other young researchers said that they had external communication much more frequently than other young colleagues in MITI-L, but that they could not help facing problems if they were to communicate with outside professionals as frequently as they liked, because such behavior might disturb 'the order and harmony of the research group.' One of them, aged 33, typically mentioned as follows.

> 'I often go out, seeking for new ideas or spurs for my research. But, generally speaking, those who like such things gradually become regarded as 'outsiders' or 'black sheep' of a group, and gradually isolated from a communication network within the group. In laboratories like MITI-L, research is actually well organized into group work, so if we were once isolated from a teamwork, it would become difficult to do what we want to, because we could not get research supports and resources enough to do that. So, we can not make the most of ideas or information from outside in the actual research work, without carefully negotiating with a boss or other colleagues. I always feel some difficulties in doing research in harmony with other people in the group.'

In short, the findings suggest that MITI-L is facing the problem of 'sunk external information,' which refers to the tendency that a large part of information obtained through external communication is inevitably left unused. In summary, there would be two interrelated reasons for the problem of 'sunk external information.' First, external communication in MITI-L is strongly biased to senior researchers representing their groups as research leaders. Nevertheless, they are not competent researchers but administrators of their group who are hardly involved in actual research activities, and consequently, less integrated into communication networks within their research group. Furthermore, since research leaders in MITI-L do not function as gatekeepers, there emerges a significant split between internal and external communication. Therefore, a large part of external information comes to be 'sunk' there at the position of a research leader. This may be the reason why external communication mysteriously appears to have a negative impact on performance. However, this would be a spurious association and, in fact, external communication must be also meaningful for researchers in

MITI-L. Second, given such a clear 'division of labor' in external communication, relatively young researchers frequently contacting information sources outside MITI-L would become isolated from leaders and other colleagues and less integrated into actual research activities within their group, since their behavior would be regarded as a disturbance to the organizational order which puts priority on teamwork. Thus, external information accessed by young researchers also tends to be sunk without being sufficiently utilized for actual research activities. After all, it might become a reasonable strategy for researchers in MITI-L to depend on internal communication and to refrain from communicating so frequently with outside professionals.

In any case, the most essential factors underlying the significant difference in performance between ERATO and MITI-L would consist in the differences in organizational mechanisms embedded in each organization rather than the apparent difference in frequency of external communication. The difference in the absolute frequency of external communication shown in Table 6.3 would relate to the difference in performance, at least to some degree. However, the analysis suggests that the relatively lower performance of MITI-L would result from its organizational mechanisms increasing 'sunk external information,' rather than merely from insufficient external communication. On the contrary, free from such impediments to full use of outside information ERATO, as the dynamic network organization, can be assumed to have advantages over bureaucratic organizations like MITI-L, not only in facilitating external communication but also in enhancing the actual utilization of external information. The essence of ERATO's organizational innovation is that it overcomes organizational mechanisms embedded in the bureaucratic organization which incur the problem of 'sunk external information.'

Implications

It has been widely recognized that one of the most serious problems in bureaucratic organizations in Japanese basic research lies in less frequent communication with outside professionals. Therefore, it has often been argued that researchers in Japanese basic research organizations should communicate with outside professionals more intensively and extensively. A number of managers in Japanese basic research organizations have also considered this as one of the most promising solutions to break the organizational deadlock. However, such arguments are potentially misleading. The thrust of findings derived from this comparative analysis emphasizes that the problems of bureaucratic organizations are more complicated and deep-rooted than generally imagined. The findings suggest that the bureaucratic organization does not only impede external professional communication, but also inevitably becomes deprived of the effectiveness of external communication itself by restraining utilization of important outside information, which results in increasing 'sunk external information.' This is not simply a matter of degree or frequency of external communication, but a more complicated structural problem resulting inevitably from organizational and managerial characteristics of the bureaucratic organization. Under such conditions within the bureaucratic organization, if managers were to facilitate external professional communication by researchers, it would only contribute to a further decline in performance. Therefore, it may be more important for Japanese basic research organizations to transform their management systems in order to introduce more organizational flexibility and fluidity, at least to some extent.

In contrast to the bureaucratic organization, the organizational experiment of ERATO suggests that the dynamic network organization could overcome such organizational mechanisms by realizing organizational flexibility and fluidity. What is important to recognize from this analysis is that the dynamic network organization could enable researchers to make full use of external information, as well as to communicate more frequently with outside professionals. In this sense, the findings in this analysis emphasize the effectiveness and the possibility of a dynamic network organization like ERATO to be more than expected. The experiment of ERATO includes an important insight for enhancing Japanese basic research activities through the creation of dynamic network organizations. Japan is now in need of increasing the number of dynamic network organizations which may strengthen basic research.

Although it is quite reasonable for bureaucratic basic research organizations in Japan to increase their organizational flexibility and fluidity by transforming their management systems, the above arguments should not be taken to mean that all bureaucratic organizations in the field of basic research are to shift to dynamic network organizations. Such an overall transformation would be neither feasible nor effective. In the Japanese context, where most of the basic research organizations are structured as stable, bureaucratic organizations, dynamic network organizations can not be entirely independent of established organizations. It is quite difficult for dynamic network organizations in Japan to recruit young, capable researchers. Thus, the organizational flexibility and fluidity in the dynamic network can be realized only on condition that there are many coexisting bureaucratic organizations with which it can interchange researchers. Moreover, bureaucratic organizations include certain kinds of advantages over dynamic network organizations such as sustainability of long-term research activities, possibility to pursue challenging or exploratory research themes, as well as easiness to accumulate relevant knowledge systematically and continuously. Since this analysis has focused on relatively short-term performance measures, it might underestimate the effectiveness of bureaucratic organizations such as MITI-L. In sum, it would become necessary to achieve and maintain a good balance between the two different types of organizations, a balance between stability and flexibility or fluidity.

One of the most important challenges for future basic research in Japan would be to establish complementary and interdependent relationships between bureaucratic and dynamic network organizations, where researchers can move from bureaucratic to dynamic network organizations, or from dynamic network to bureaucratic organizations, while either type typically pursues its advantages at the same time. Such complementary relationships between bureaucratic and dynamic network organizations suggest the possibility of a totally new inter-organizational matrix structure in the field of basic research in Japan. The creation of the dynamic network organization in the Japanese context also signifies the potential of Japanese basic research to generate original social systems.

APPENDIX

Table 6.7 Factor analysis of communication variables

Variables	*F1*	*F2*	*F3*	*F4*	*F5*
4 Professional communication with researchers outside of the organization (the same specialty)	0.87	0.12	0.04	0.23	0.07
5 Professional communication with researchers outside of the organization (different specialties)	0.84	0.10	0.09	0.25	0.06
6 Attendance at academic meetings	0.81	−0.05	−0.12	−0.07	−0.14
1 Professional communication with researchers within the organization (the same group)	0.06	0.77	0.21	−0.21	0.18
2 Professional communication with researchers within the organization (different groups)	−0.18	0.76	−0.39	−0.04	0.01
3 Professional communication with supervising researchers	0.35	0.64	0.15	0.10	0.01
8 Lunch with colleagues	−0.03	0.07	0.93	0.05	−0.05
9 Engagement in joint research	0.22	−0.08	0.05	0.93	−0.08
7 Attendance at formal meetings	−0.01	0.13	−0.05	−0.07	0.97
Eigenvalues	2.34	1.64	1.13	1.04	1.02
Percentages of variance explained	29.4%	20.1%	12.6%	9.6%	8.1%
Cumulative percentages	29.4%	49.4%	62.1%	71.7%	79.8%

Source: Author's composition.

Notes:

a The principal component analysis is used. The numbers are factor loadings after Varimax rotation.

b F stands for Factor.

REFERENCES

Allen, Thomas J. (1977) *Managing the flow of technology.* Cambridge: MIT Press.

Allen, Thomas J. and Steven Cohen (1969) Information flow in R&D laboratories. In *Administrative Science Quarterly*, vol. 14, pp. 12–19.

Allen, Thomas J., Michael Tushman, and Denis Lee (1979) Technology transfer as a function of position on research, development, and technical service continuum. In *Academy of Management Journal*, vol. 22, pp. 694–708.

Eisenhardt, Kathleen M. (1989) Building theories from case study research. In *Academy of Management Review*, vol. 14, pp. 532–50.

Fielding, Nigel G. and Jane L. Fielding (1986) *Linking Data*. California: Sage.

Glaser, B. G. and A. L. Strauss (1967) *The Discovery of Grounded Theory: Strategies for Qualitative Research.* New York: Aldine Publishing.

Hagstrom, Warren (1965) *The Scientific Community.* New York: Basic Books.

Iinuma, Kazumasa, Mitsuo Iinuma, and Ken Kusunoki (1991) *Kisōteki sendōteki kagaku gijutsu no suishin no tame no kenkyū jinzai ni kansuru chōsa kenkyū* (A study on human resource management for basic research and advanced engineering). Tokyo: Institute of Future Technology.

Kanai, Toshihiro (1989) *Entrepreneurial Networking: A Comparative Analysis of Networking Organizations and their Participants in an Entrepreneurial Community.* PhD dissertation, Sloan School of Management, MIT.

Katz, Ralph (1978) Job longevity as a situational factor in job satisfaction. In *Administrative Science Quarterly*, vol. 23, pp. 204–23.

—— (1980) Time and work: Toward an integrative perspective. In Staw, B. and L. L. Cummings (eds), *Research in Organizational Behavior*, vol. 2, pp. 81–127. Connecticut: JAI Press.

—— (1982) The effects of group longevity on project communication and performance. In *Administrative Science Quarterly*, vol. 27, pp. 81–104.

Katz, Ralph and Michael Tushman (1979) Communication patterns, project performance and task characteristics: An empirical evaluation and integration in an R&D setting. In *Organizational Behavior and Human Performance*, vol. 23, pp.139–62.

Katz, Ralph and Thomas J. Allen (1982) Investigating the Not Invented Here (NIH) syndrome: A look at the performance, tenure, and communication patterns of 50 R&D project groups. In *R&D Management*, vol. 12, pp. 7–19.

Mansfield, E., and S. Wagner (1975) Organizational and strategic factors associated with probabilities of successful industrial research. In *Journal of Business*, Winter, pp. 179–98.

Okimoto, Daniel (1986) The Japanese challenge in high technology. In Landau, Ralph and Nathan Rosenberg (eds), *The Positive Sum Strategy: Harnessing Technology for Economic Growth.* Washington, DC: National Academy Press, pp. 541–68.

Pelz, Donald, and Frank M. Andrews (1976) *Scientists in Organizations: Productive Climate for Research and Development* (rev. edn). Michigan: Institute for Social Research.

Pfeffer, Jeffrey (1983) *Power in Organizations.* Massachussets: Pitman.

Porter, Michael E. (1990) *The Competitive Advantage of Nations.* New York: Free Press.

Sakakibara, Kiyonori (1991) Increasing basic research in Japan: Corporate activity alone is not enough. In *Current Politics and Economics of Japan*, vol. 1, pp. 83–8.

Smith, Clagett G. (1979) Age of R&D group: A reconsideration. In *Human Relations*, vol. 23, pp. 81–93.

Thompson, James D. (1967) *Organization in Action*, New York: McGraw-Hill.

Tushman, Michael (1979) Managing communication networks in R&D laboratories. In *Sloan Management Review*, vol. 20, pp. 37–49.

Tushman, Michael and Ralph Katz (1980) External communication and project performance: An investigation into the role of gatekeepers. In *Management Science*, vol. 26, pp. 1071–85.
Yin, Robert K. (1984) *Case Study Research: Design and Method*. California: Sage.

7

THE JAPANESE BUSINESS SYSTEM FOR CREATION AND DIFFUSION OF TECHNOLOGICAL KNOWLEDGE

Time for change?

Hiroyuki Odagiri

Introduction: the nature of technology

It is well-known that technology possesses characteristics akin to public goods, such as non-rivalry and non-excludability. Non-rivalry means that my use of a certain technological knowledge need not preclude you from using exactly the same knowledge, in contrast to private goods in which my eating an apple for instance automatically precludes you from eating the same piece of apple. Non-excludability means that I cannot exclude you from using the same knowledge. Patents, copyrights, and other intellectual property rights are artificial devices to create exclusivity to intrinsically non-excludable knowledge. Yet they are not complete and the resulting lack of appropriability, that is, of a mechanism for an inventor to appropriate all the returns from invention, has been argued to cause under-investment in technology (Arrow 1962).

From a social viewpoint, however, it is desirable to diffuse a knowledge widely because, once invented, the marginal cost of diffusion is supposedly small. A dilemma is therefore inevitable: on the one hand, creating appropriability is desirable to enhance incentives for invention; on the other, encouraging diffusion is desirable from the viewpoint of a socially efficient use of invention.

Yet, the following may be stated. First, many firms actually do not rely on patents to appropriate returns from their inventions.[1] Second, whether patent protection is effective or not, the firm cannot survive without innovation efforts, particularly in high-technology industries. To this extent, inappropriability of technology as a deterrence to invention seems to have been overstated. Third, the diffusion of technological knowledge should be fostered so long as it does not conflict with innovation incentives. For instance, in developing economies where imported technology is an important source of economic development, a widespread and effective use of the technology can be a key determinant of the speed of its development.

However, the assimilation of imported (or imitated) technology is by no means easy mainly because, unlike normal goods, technology is not something just to be bought and consumed. An individual, a firm, or a country that imports technology has to possess capabilities to understand it and apply it in its own environment which often differs significantly from the environment in which the technology was invented. Furthermore, the extent and content of capabilities one has at a particular point in time depends on the path in which one has accumulated capabilities. This path-dependency is another salient characteristic of technology and is increasingly recognized as the factor to be reckoned with in discussing the growth of the firm or of the country (see, for instance, David 1985 and Teece 1988). It is a characteristic of technology in an inter-temporal dimension whereas inappropriability is a characteristic in a spatial dimension.

In this paper, I intend to discuss the implications of these characteristics of technology for a theory of the firm and economic growth, and how they should be taken into account in understanding the technological progress in Japan today and in the past. These discussions hopefully will shed light in understanding the changes that may (or should) take place in the Japanese business and industrial system and their consequence to technical progress.

The development of technological capabilities in Japan

In *Technology and Industrial Development in Japan*, Odagiri and Goto (1996) used historical case studies to show how the Japanese industries developed their technological capabilities. These case studies clearly indicated that importation of technology was by no means a simple matter. First of all, in a yet undeveloped economy, importation of technology is both difficult and risky. A strong entrepreneurship is required in the sense of both Schumpeter (1942) who stressed the willingness to strive to gain something new and unfamiliar, and Knight (1921) who stressed the willingness to bear risk. For example, when E. Shibusawa and T. Yamanobe established Japan's first large-scale cotton-spinning mill in Osaka, the amount of investment was huge. Even though they imported all the machines from England, the choice of location, machine installation and the training of workers to handle the machines required many difficult decisions and hardship. To acquire the basic technological knowledge, Yamanobe spent two years as an apprentice in England, where he met a series of difficulties beginning with the refusal of all the mills there, except one, to take him as an apprentice.

Although they hired an English engineer (at a surprisingly high salary) to supervise the installation of the spinning machines imported from his company, his knowledge extended little beyond them. Therefore, Yamanobe had to struggle to learn how to install and operate other machines, such as steam engines. Remarkably, Yamanobe and other Japanese engineers had the capacity to understand the manual and apply it, and, at least in one case, even to find a mistake in the manual and correct it. The scientific level during the preceding Tokugawa era (1600–1867) had been significantly high despite the Shogunate government's restriction of inflow of Western knowledge. Furthermore, elementary education had been probably more widespread, mostly as private efforts, than in Europe and America at the time. Adding to these backgrounds, the government after the Meiji Restoration (1868) hurried to establish a nation-wide elementary-education system and higher-education institutions, including an engineering school.

It was this combination of entrepreneurship and the presence of capabilities thus nurtured that made the rapid development of major industries possible in Japan. Importation of technology and catch-up were, therefore, not at all a simple matter of buying a blue print and an advanced machine. A high level of capabilities were actually needed to assimilate the imported technology.

To give a more recent example, when Sony and other firms began their efforts to manufacture transistors after buying the necessary patent from the US, not only scientific and technological understanding, but also a sufficient level of know-how in production management and labor skills, was needed. The supply of materials and components was another obstacle because, for instance, sufficiently pure silicon (or even the knowledge that silicon can be made so pure) was lacking and, as a result, all the early experiments failed. Yet, the firms and their researchers kept experimenting and attempted to mass-produce transistors when none of the American firms believed mass production was commercially viable.

This example teaches us that one has to adapt and modify foreign technologies to fit them to one's own natural and economic needs (also see Rosenberg 1976). To take yet another example, when Japan imported German engineers and German state-of-the-art steel-making facilities including furnaces around the turn of the century, these engineers did not realize that coal could be of different properties between the two countries. This ignorance, among others, caused the operation to terminate after a year of continuing difficulties. It was a Japanese engineer who determined the cause and corrected the problem by, for instance, re-designing the imported furnaces.

A similarly high level of capability was needed to select appropriate technologies. Such selection can only be made with a sufficient understanding of both the technologies themselves and the environment in which they are to be utilized. Furthermore, the selection can be risky because, particularly at an early stage of technology, its potentials and limitations are as yet unknown.

How can the capabilities be nurtured and accumulated to select, absorb, and adapt imported technologies and, eventually, to start inventing new technologies of its own? Needless to say, education and training are essential in enhancing the general ability. Still, they are not sufficient because the process of trial and error, including learning, can be equally indispensable. Unless one really experiences a technology, one would never understand the potentials as well as the limitations. And, without such understanding, one would never be able to improve it or start new innovation. Furthermore, learning has to take place at any level of the firm from managers and engineers to shop-floor workers. Managers, in other words, have to provide the workers with opportunities to learn from the new technology. It can be risky, particularly if the new technology has to be used at the site of actual production rather than within laboratories. In several cases, including Kawasaki Steel's introduction of continuous casting and Sony's development of a transistor production system, such efforts and risk were indeed taken by Japanese entrepreneurs and managers (Odagiri and Goto 1996).

Learning by doing was an important way of increasing productivity; yet, learning was never an automatic process. The popular concept of learning as a function of accumulated volume of production is misleading: one has to be alert to what the experience teaches. One has to possess an ability to notice and understand what the experience infers. Similarly, the diffusion or spillover of technology is not an automatic process either. One has to be alert to whatever information may come from others and one often has to make conscious efforts to learn what others are doing. You cannot just sit at a lecture to learn: you have to be alert to what the lecturer is saying and have sufficient ability to understand it.

The speed of learning and diffusion also depends on industrial and business organization, because a certain organizational form may be better suited for learning than others. This topic is discussed in the next section.

All the discussions in this section point to inter-temporal dependence of technology. How well the firm or the country can assimilate a new technology or how much it can invent depends on the technological capabilities it possesses at the time, which in turn are influenced by the past experience and efforts of the firm (or the country). Path-dependencies are the very characteristics of technology that are bound to influence the strategy of the firm.

Technology and the business system

Much of technological knowledge is tacit by nature. Certainly, some types of knowledge can be written on paper (or in an electronic file) and mostly understood by those who read it. However, a wide class of knowledge can be only partially transmitted by paper, file, or any other communication method. Moreover, written information may be fully exploited only with associated tacit knowledge, often called know-how. Economics professors usually find it easy to teach an economics theorem but find it extremely difficult to teach how the theorem can be applied to real economic phenomena. Only with the student's persistent effort to apply the theorem, occasionally with silly failures and misunderstanding, he/she would start to understand the power and limitations of the theorem and gain the know-how in the use of it.

Furthermore, know-how is often specific to the organization or to the environment. One's know-how may not be useful in another organization and it may become effective only in combination with the know-how of other members of the same organization. This is particularly true with the firm where team production is taking place, because know-how and skills are more appropriate within the team than outside. The term 'firm-specific skills' refers to such skills. However, skills here should be interpreted in a broad context. They refer not only to the usual sense of skills, say, skills in maneuvering machines and equipment, but generally to various kinds of tacit knowledge including, for instance, the knowledge (or, more appropriately, competence) of how to fill in particular forms of paper, how to deal with emergencies, how to build good relationships with partners and customers, and how (and where) to apply new knowledge to improve productivity.

Firm-specificity of these skills have been the prime reason why many functions of labor markets have been internalized, that is, carried out within the firm (Doeringer and Piore 1972, Williamson 1975). Long-term employment relationships are common not only in Japan where the practice (particularly among large firms) of so-called lifetime employment has been well documented, but also in the US and most industrialized nations (Odagiri 1992). Internal promotion with occasional transfer across jobs or work places is a norm rather than an exception. These practices have been conducive to workers' accumulating skills, specific or general, through many years of experience, on-the-job training, job rotation, and perhaps outside education. An internal labor system, therefore, is a rational device to deal with the need to maintain and foster competence within the firm. By no means was it a consequence of paternalism. In fact, the Japanese experience suggests that such an internal labor system gradually developed during the First World War boom and the following years when the shortage of skilled workers and volatile movement of workers was a serious impediment to smooth plant operation and further expansion of the firm (Odagiri and Goto 1996).

The firm, as a result, may be viewed as a collection of skills (in a broad sense as discussed earlier) of its members, and these skills give the firm its unique capabilities. The point is that as an organization, the firm can attain capabilities that are more than a mere summation of skills of individual members because their skills can be fully productive only within the context of the organization they have been in. Chandler (1990) thus used the words 'organizational capabilities' to discuss the development of major firms in the US, UK and Germany.

The presence of and the necessity for these capabilities constrain the firm in its growth strategy. First, growth of the firm has to be associated with the growth of capabilities. Whereas the expansion of physical facilities may be easy as long as the supply of equipment and construction work abounds, the expansion of capabilities, particularly the expansion of the workers with necessary skills, can be much more difficult. Consequently, the faster the firm intends to grow, the more inexperienced and less talented the new workers are bound to be, impeding growth or causing additional costs. This fact was emphasized by Penrose (1959), who gave a foundation to the theory of growth of the firm (Odagiri 1981).

Second, growth should be pursued in a direction where the firm has a comparative advantage, that is, where its core capabilities or core competencies can be fully exploited (Prahalad and Hamel 1990). This fact clearly implies that diversification to related fields is more advantageous than pure conglomerate-type diversification, given other conditions such as the growth potential of various industries. In fact, most cases of diversification indicate that the coherence across diversified activities is maintained (Teece, Rumelt, Dosi, and Winter 1994). This is particularly true when diversification is attained internally. When it is made through mergers and acquisitions, there is an increased tendency to diversify toward less related fields because the capabilities in these fields may be gained together with the acquired firm. Yet, the management has to possess familiarity with these acquired capabilities, suggesting again the disadvantage of pure conglomerate mergers.

The advance of our understanding of the nature of the firm, particularly in relation to technology, along the lines discussed so far owes greatly to such authors as Chandler, Nelson, and Teece.[2] The discussion also suggests that many phenomena attributed to Japanese firms have been (and probably should be) actually applied by other firms universally. These include the internal labor system as discussed earlier, the emphasis on on-the-job training with occasional rotation, the preference of internal growth over mergers and acquisitions, and limited diversification to unrelated fields (Odagiri 1992). The long-term relationship with the suppliers and customers, usually called the *keiretsu* relationship, is another. Such relationship is necessary because in many cases such as automobile assembly, both the suppliers and the assemblers have to accumulate relation-specific skills (Asanuma 1989).

The long-term relationship is important in relation also to the spatial nature of technology discussed earlier. Because technological knowledge tends to spill over through transactions, the maintenance of long-term relationships is helpful in preventing knowledge from leaking out of the group. In this context, the *keiretsu* group may be considered a device to internalize the spillover. When, for instance, an automobile assembler develops a new model, it needs to order a new, specific part from a supplier. If it has to worry that the design and specification of the part may become known to rival assemblers, the incentive for product innovation will be lessened or it will have to integrate vertically.

The important concept here is reputation. By being loyal to the assembler, that is, by collaborating with the assembler and not leaking the information to the outside, the supplier creates a reputation, thereby gaining trust from the assembler. The assembler, on the other hand, creates a reputation by not abusing the supplier. Mutual trust is thus created, making the quasi-internalization of knowledge possible. 'Prisoners dilemma' is thus resolved, if we use the terminology of game theory (Kreps 1990).

This quasi-internalization provides a perfect setting for collaborative research and development. The participants to the collaboration can contribute their complementary technological capabilities without a need to worry that the partner might betray or steal their knowledge. Indeed, a survey by the Fair Trade Commission of Japan (Rokuhara 1985) concerning electric and communications equipment, automobile and parts, chemicals, and other materials industries indicate that the majority of R&D collaborations have been made between firms in vertical relationships. As a reason for conducting joint R&D, the largest proportion of the participating firms cited the augmentation of own technologies with those of the partners. Furthermore, the majority of joint R&D agreements included the restriction (or prohibition) on the supply of technology to third parties or on joint R&D with third parties. These facts clearly indicate that the combination of complementary technological capabilities is the major motivation for collaborative R&D but the firms are deeply concerned with the leakage of technologies to the outside, thus creating a dilemma. The mutual trust created through long-term relationships greatly helps to resolve this dilemma. In this regard, the Japanese *keiretsu* system has played an important role. It is no wonder that the Japanese auto-makers have performed better in product development than American and European ones and that Americans have been catching up by imitating the Japanese method (Fujimoto 1997).

Estimated intra- and inter-industry spillovers

The discussion in the previous section suggests that substantial technological spillovers take place within *keiretsu*. Suzuki (1993) estimated for the Japanese electrical machinery industry that both R&D spillovers among core firms and the spillovers (or technology transfer as the author calls it) from core firms to their subcontracting firms are significant. According to his estimate, the spillovers add 3.6 percent and the transfer adds 8 percent to the private return (20.2 percent) of R&D capital of the core firms.

The estimation of intra- and inter-industry R&D spillovers has been made by several authors, mostly in the US (refer to surveys by Griliches 1992 and

Nadiri 1993). Most commonly, their studies estimated production functions or cost functions that include the R&D stock variable(s) of other industries in addition to conventional production factors (or their prices) and own R&D stock.

Presumably, these spillovers take place through two channels. One occurs at the R&D stage where researchers may obtain new knowledge from journals and presentations at conferences, which can be applied to their own research. They may learn that a certain research is in progress in another company and ask for the latter's collaboration to solve bottlenecks in their R&D. The likelihood for such spillovers, diffusion, or collaboration must be higher the nearer the research subjects of the two parties are (and lower if the two parties compete against each other in markets because, if they do expect to be rivals, they would be unwilling to share the knowledge). Thus, this likelihood is measurable by the similarity of the distribution of R&D expenditure across research fields, termed R&D proximity (Jaffe 1986).

The other is the spillover that takes place when intermediate and capital goods are purchased from other industries. When the capacity or other quality measures of a machine are increased through product development, the industry using the machine will be able to increase the productivity and reduce costs.[3] The spillovers occurring in this fashion are called R&D flow and are usually measured by the R&D expenditure of the source industry multiplied by the proportion of this industry's sales to the industry receiving the spillover (Terleckyj 1980).

Odagiri and Kinukawa (1997) calculated the R&D proximity and R&D flow for four industries in Japan – general machinery, electrical machinery, transportation machinery and chemicals – and confirmed that some of the spillovers that are expected to exist through either of these two channels do in fact reduce the cost of the industry receiving them. These include spillovers from metal products to general machinery and transportation machinery, and from chemicals to electrical machinery. No source industry was found to have such an effect upon chemicals.

The chemicals industry has a high R&D flow to electrical machinery, suggesting that the cost-reducing effect takes place through the latter's purchase of chemical products, say, silicon, as intermediate goods. By contrast, the metal products industry has a high R&D proximity with both general machinery and transportation machinery, suggesting that the spillovers at the stage of R&D were important. These results indicate that spillovers are actually effective and, furthermore, that the channels of spillovers can be various.

Admittedly, the methodology for the estimation of spillover effects is still primitive. This is particularly true in view of the discussion we have made in previous sections. An industry may gain from the technologies in other industries in both explicit and implicit fashions because some of them are tacit, whereas all the econometric studies have been relying on R&D

expenditure as measured in official statistics. If we were able to measure the spillover of tacit knowledge as well, the effects might well have been greater than what the above estimates indicated.

The presence of these spillovers implies that the social rate of return to R&D investment exceeds the private rate. As discussed earlier in relation to inappropriability, this fact implies on the one hand that the private incentive for R&D may be smaller than is socially desirable. On the other hand, provided that the private incentive is not impaired, it is socially needed to maximize the gains from spillovers. For instance, the positive spillover effect from chemicals to electronics may have been caused by the improved quality of silicon wafers produced by chemical firms, which improved the competitiveness of semiconductors produced by Japanese electronics firms. The dynamic development of an industry may thus crucially depend on the development in associated industries. Such was in fact the case when many Japanese industries started their development such as shipbuilding, automobiles, and semiconductors (Odagiri and Goto 1996). Without technically competent suppliers available, the firms had to nurture suppliers or collaborate with other firms to jointly develop necessary material and components. Such collaboration has developed into the present *keiretsu* system.

In several cases, collaboration also took place horizontally, that is, among the firms in the same industry. In a few cases, most notably computers, such collaboration was undertaken with governmental assistance. Research collaborations founded on this basis have been well documented elsewhere (Goto 1997). Thus, we emphasize here just three facts. First, these collaborations were not always successful. Second, there is little evidence that they fostered collusive behaviour in the product markets. In fact, by making the technological level of participating firms comparable, they may have enhanced competition. Finally, emphasis in joint R&D has shifted recently toward research in a more basic nature, for instance, the Fifth-Generation computer project (Odagiri, Nakamura and Shibuya 1997). This last tendency also implies that, even for basic research, the Japanese innovation system tends to rely on the private business sector more than other advanced nations.

Time for change?

In recent years, particularly after the collapse in 1991 of the so-called bubble economy, many observers argue that the Japanese business system has started to change and/or should change. Let me first list some of these arguments:[4]

1 The Japanese system may have been effective in fostering incremental innovation; however, Japan now needs more drastic innovation. The system that emphasizes teamwork and cooperation is ill-suited for the promotion of individual originality.

2 The system may have been well-suited in assembly industries; however, it is now going to be the age of software and information, in which venture businesses – Microsofts, Netscapes, and the like – play significant roles. The Japanese system does not provide support to these businesses.
3 The Japanese company is now suffering from the inflexibility caused by its lifetime employment system. It should have freer hands in restructuring its workforce by eliminating excessive (and obsolete) workers and by recruiting workers with suitable skills in mid-career.
4 The aging of the workforce has increased labor costs because of the seniority-based wage system. Wages should be based more on merit rather than seniority.
5 The share-ownership structure of Japanese firms, characterized by majority ownership by financial institutions and friendly companies, resulted in the lack of governance. This lack is responsible for the post-bubble failure of a number of companies (including financial institutions).
6 The rapidly expanding foreign direct investment of Japanese firms and the transfer of production capacity abroad have resulted in the hollowing-out of Japanese manufacturing, causing the loss of skills in large firms, the collapse of small firms with specialized skills, and the difficulty in maintaining smooth manufacturing-R&D linkages.

Doubtless these arguments capture certain important aspects of the limitation of the Japanese business system. It is also true that a changing economic and technological environment has rendered obsolete some of the once-effective business practices. Particularly, the virtual end of the catch-up era has increased the need for original research and, relatively, diminished the efficiency of the internal labor system. Yet, the discussion of the nature of technology in previous sections is still relevant. What I called the spatial nature (spillovers) and inter-temporal nature (path-dependencies) are the very basic characteristics of technology and apply to radical as well as incremental innovation to varying degrees. A large, and probably essential, part of the Japanese business system is a device, gradually built and adapted, to deal with these natures. A process of selection in the evolutionary sense existed, leaving the better suited and gradually replacing the obsolete. Learning of others' better systems took place. It is no wonder that, as discussed already, many practices are actually found in other countries as well.

It is also important to note that some of the above discussions are based on too stereotyped views of the Japanese system. For example, venture businesses have never been absent in Japan. Many well-known companies today started more or less as venture businesses, from Toshiba and Toyota to Sony and Honda (Odagiri and Goto 1996), and the proportion of medium and small firms has been (and is) larger than in other countries.[5] Even though the activity of venture businesses in electronics, software, and biotechnology in the US has been conspicuous, many small companies in Japan have been innovative and growing in various industries.

In fact, in a recent visit to several mid-sized Japanese firms ranging from manufacturing of pumps and lathe chucks to furniture and school uniforms, the author found them to have a surprisingly high level of technology and to be still actively investing in innovation. The words, 'technology' and 'innovation,' should be taken in a broad sense just as Schumpeter implied. Prompt or punctual delivery and after-sales service may be as important as good quality and newness of the product, and maintaining an organization that is capable of accomplishing them is an innovation. This is exactly why several writers prefer to depict the competitiveness of the firm by such words as 'organizational capabilities' (Chandler 1990) and 'core competence' (Prahalad and Hamel 1990). Too much emphasis on the so-called high-tech industries and on narrowly defined science and technology may actually impair the competencies of the economy. Even if (which is a big 'if') the Japanese business system is not suitable for original and basic research, its complete abandonment may harm the competencies in other areas and, in balance, Japan may suffer. Changes will have to be made carefully so that Japan's comparative advantage should be preserved: indeed, the evolutionary process of natural selection is likely to lead the change in that direction.

Another important issue is the globalization of business activity. Globalization has affected the Japanese business system in two ways. First, as discussed in point 6 above, the loss of some of the manufacturing activities within the firm (or by the suppliers) has caused a loss of accumulated skills and a difficulty in maintaining feedback from manufacturing to R&D. There seems a real danger here particularly because path-dependencies suggest that, once lost, it can be costly and time-consuming to regain the technology. Many large Japanese firms appear to be aware of this danger and have been maintaining the advanced part of manufacturing in Japan however costly it is relative to the production in low-wage countries. In this regard, the 'long-termism' of Japanese firms, as exemplified by their pursuit of growth, may have helped in lessening the impact of hollowing-out; and the ownership structure as contained in point 5 above may have been in fact contributing (Odagiri 1992).

Second, as the firms became multinational, a more transparent and more internationally-applicable management system has become necessary. They now employ foreign directors, foreign management staff, and foreign researchers: accordingly personnel policy and supervision have to be modified to accommodate them. In addition, international harmonization of accounting, labor, patents, and other legal systems is in progress, which may force the firms to accept a more transparent, rule-based system. These changes, however, need not be in conflict with the core of the Japanese business system as discussed earlier, such as long-term employment and other business relations; the emphasis on intrafirm accumulation of skills through training and learning, the high value placed on reputation and trust, and the coherence of business activities. Indeed, the design and implementation of a new business system that is more internationally-oriented and yet preserves the better part of the so-called Japanese system seems to be the biggest task for Japanese firms. Only those firms, in Japan or elsewhere, succeeding in this innovation may be able to survive the world-wide evolutionary selection process in coming years.

NOTES

1 For American results, see Levin, Klevorick, Nelson, and Winter (1987). They indicate that, for many firms, lead time and learning-curve advantages are more important means of appropriation. A similar finding was made for Japanese firms in a yet unpublished study conducted by the National Institute of Science and Technology Policy led by Akira Goto.

2 In addition to the literature cited already, a useful reference are the papers in the Winter special issue of *Strategic Management Journal* (vol. 12, 1991).

3 Actually, if the price of the machine also increases reflecting fully the increase in quality, the cost function would be unaffected. However, to the extent that the price increase is not large enough to offset the quality increase, the cost-reducing effect of R&D spillovers would take place.

4 These arguments have been usually put forward in newspapers and other popular writings rather than in academic articles: therefore, no reference will be given. For discussions surrounding the hollowing-out issue, see Nakamura and Shibuya (1995).

5 The proportion of workers employed in firms with less than 100 employees was 55 percent in Japan in 1990 (Ministry of International Trade and Industry, Census of Manufactures) and 30 percent in the US in 1987 (Department of Commerce, Census of Manufactures).

REFERENCES

Arrow, Kenneth J. (1962) Economic Welfare and the Allocation of Resources for Invention. In National Bureau of Economic Research, *The Rate and Direction of Inventive Activity*. Princeton, NJ: Princeton University Press, pp. 609–25.

Asanuma, Banri (1989) Manufacturer-Supplier Relationships in Japan and the Concept of Relation-Specific Skills. In *Journal of the Japanese and International Economies*, vol. 3, pp. 1–30.

Chandler, Alfred D., Jr. (1990) *Scale and Scope*. Cambridge, Massachusetts: Harvard University Press.

David, Paul A. (1985) Clio and the Economics of QWERTY. In *American Economic Review*, vol. 75, pp. 332–7.

Doeringer, Peter B. and Michael J. Piore (1972) *Internal Labor Markets and Manpower Analysis*. Lexington, Massachusetts: Heath Lexington Books.

Fujimoto, Takahiro (1997) The Dynamic Aspect of Product Development Capabilities: An International Comparison in the Automobile Industry. In Goto, Akira and Hiroyuki Odagiri (eds), *Innovation in Japan*. Oxford: Oxford University Press.

Goto, Akira (1997) Co-operative Research in Japanese Manufacturing Industries. In Goto, Akira and Hiroyuki Odagiri (eds), *Innovation in Japan*. Oxford: Oxford University Press.

Griliches, Zvi (1992) The Search for R&D Spillovers. In *Scandinavian Journal of Economics*, vol. 94, Supplement, pp. 29–47.

Jaffe, Adam B. (1986) Technological Opportunity and Spillovers of R&D: Evidence from Firms, Patents, Profits, and Market Value. In *American Economic Review*, vol. 76, pp. 984–1001.

Knight, Frank H. (1921) *Risk, Uncertainty and Profit*. Boston, Mass.: Houghton Mifflin. Reprinted: Chicago, Illinois: The University of Chicago Press, 1971.

Kreps, David M. (1990) Corporate Culture and Economic Theory. In Alt, James E. and Kenneth A. Shepsle (eds), *Perspectives on Positive Political Economy*. Cambridge: Cambridge University Press, pp. 90–143.

Levin, Richard C., Alvin K. Klevorick, Richard C. Nelson and Sidney G. Winter (1987) Appropriating the Returns from Industrial Research and Development. In *Brookings Papers on Economic Activity*, vol. 3, pp. 783–820.

Nadiri, M. Ishaq (1993) *Innovations and Technological Spillovers*. Washington, DC: National Bureau of Economic Research (Working Paper no. 4423).

Nakamura, Yoshiaki and Minoru Shibuya (1995) *The Hollowing Out Phenomenon in the Japanese Industry*. Tokyo: Ministry of International Trade and Industry, Research Institute of International Trade and Industry (Studies in International Trade and Industry, no. 19).

Odagiri, Hiroyuki (1981) *The Theory of Growth in a Corporate Economy: Management Preference, Research and Development, and Economic Growth*. Cambridge: Cambridge University Press.

—— (1992) *Growth through Competition, Competition through Growth: Strategic Management and the Economy in Japan*. Oxford: Clarendon Press.

Odagiri, Hiroyuki, Yoshiaki Nakamura and Minoru Shibuya (1997) *Research Consortium as a Vehicle for Basic Research: The Case of a Fifth Generation Computer Project in Japan*. In *Research Policy*, vol. 26, pp. 191–207.

Odagiri, Hiroyuki and Akira Goto (1996) *Technology and Industrial Development in Japan: Building Capabilities by Learning, Innovation, and Public Policy*. Oxford: Clarendon Press.

Odagiri, Hiroyuki and Shin'ya Kinukawa (1997) *The Contributions and Channels of Inter-Industry R&D Spillovers: An Estimation for Japanese High-Tech Industries*. In *Economic Systems Research*, vol. 9, pp. 127–42.

Penrose, Edith T. (1959) *The Theory of the Growth of the Firm*. Oxford: Basil Blackwell.

Prahalad, C. K. and Gary Hamel (1990) The Core Competence of the Corporation. In *Harvard Business Review*, May–June, pp. 79–91.

Rokuhara, Akira (1985) *Kenkyū kaihatsu to dokusen kinshi seisaku* (R&D and Antimonopoly Policy). Tokyo: Gyosei.

Rosenberg, Nathan (1976) *Perspectives on Technology*. Cambridge: Cambridge University Press.

Schumpeter, Joseph A. (1942) *Capitalism, Socialism, and Democracy*. New York: Harper and Row.

Suzuki, Kazuyuki (1993) R&D Spillovers and Technology Transfer among and within Vertical Keiretsu Groups: Evidence from the Japanese Electrical

Machinery Industry. In *International Journal of Industrial Organization*, vol. 11, pp. 451–602.

Teece, David J. (1988) Technological Change and the Nature of the Firm. In Dosi, G., C. Freeman, R. Nelson, G. Silverberg and L. Soete (eds), *Technical Change and Economic Theory*. London: Pinter, pp. 256–81.

Teece, David J., Richard Rumelt, Giovanni Dosi, and Sidney Winter (1994) Understanding Corporate Coherence: Theory and Evidence. In *Journal of Economic Behavior and Organization*, vol. 23, pp. 1–30.

Terleckyj, Nestor E. (1980) Direct and Indirect Effects of Industrial Research and Development of the Productivity Growth of Industries. In Kendrick, J. W. and B. N. Vaccara (eds), *New Developments in Productivity Measurement and Analysis*. Chicago: The University of Chicago Press, pp. 359–77.

Williamson, Oliver E. (1975) *Markets and Hierarchies*. New York: Free Press.

8

REORGANIZATION OF R&D IN JAPANESE MANUFACTURING FIRMS

Preserving competitiveness for the twenty-first century

Martin Hemmert

Introduction: challenges for Japanese industrial R&D in the 1990s

Japan has stunned the world over decades with its successful economic catch-up process to the Western industrialized countries. A prerequisite for this was the technological catch-up. In order to improve the technological level of the Japanese economy, equipment and know-how was imported through several channels and by various means. This meant not simply the imitation of existing technology, but required the ability to adapt it to the Japanese environment and to distribute it among individuals and institutions. As a consequence, a sophisticated framework of technology transfer and technology utilization developed (Francks 1992; Odagiri and Goto 1993; Odagiri and Goto 1996). The profound success of these continued efforts is best indicated by the fact that Japanese companies in the 1980s took the lead on world markets in several fields of high technology.

As in many cases, however, the success itself finally tends to render obsolete the system it was based on. The very fact that Japanese firms have reached world level in most technologies relevant for industrial manufacturing also implies that the potential for technology transfer from abroad has now become very limited, or, in some areas, almost nonexistent. Therefore, the need to strengthen the domestic system of technology generation is felt stronger than ever in the 1990s. Policy is already changing in order to redirect the Japanese innovation system: public investment in R&D is significantly increasing, venture capital programs for innovative small and medium enterprises are being launched, and universities are striving to increase capacity for postgraduate studies.

The bulk of the Japanese R&D effort however is still financed and performed by private firms (Somucho Tokeikyoku 1997: 6). The following paragraphs focus on the reorganization of R&D in Japanese manufacturing firms which account for more than 93 percent of the R&D expenses in the corporate sector (Somucho Tokeikyoku 1997: 15). What are the main challenges for Japanese industrial R&D in the 1990s? Two issues are frequently discussed in this context.

First, as already mentioned above, there is a vivid debate about the necessity of strategic reorientation of Japanese R&D that also covers the corporate sector. Strong concern is expressed about the present structure of industrial R&D, indicating that while the system of technology diffusion and technology utilization is well developed, there are fundamental weaknesses in technology generation. Therefore, it is argued, increased efforts in research, underpinned by a higher organizational independence of R&D units and more individual-based incentives for R&D workers, are needed (Akashi 1995: 20–2; Sakakibara 1995: 270–2; Seisansei Kenkyusho 1996). This leads to the question to what extent the companies are actually changing their R&D management in that direction.

There is, however, a second important factor that may have seriously affected the direction of Japanese industrial R&D in recent years. Since 1991, a severe and long-lasting recession struck the Japanese economy. In particular, the manufacturing sector was suffering from the concurrent loss of international competitiveness because of the Yen evaluation and the depression of domestic demand. Under such circumstances, management is under strong pressure to reduce costs and to improve profitability in the short and medium term. Moreover, activities like R&D that are not directly linked to sales performance tend to come under especially strong cost reduction pressure. The need for rationalization has a shorter time horizon and is therefore on a somewhat different dimension than the above-mentioned strategic reorientation. Nevertheless, the impacts of these two external factors on R&D management are fundamentally opposed, as the first one calls basically for expansion, the second one for retrenchment.

This raises the question as to which of the two challenges has been given priority in the managerial strategies of Japanese manufacturing firms in recent years. Are the firms primarily striving for the strengthening of their R&D abilities or for cost reduction in this field in order to improve short and medium term profitability? Are there attempts to pursue both strategies simultaneously?

These issues will be examined in the following chapters. The paper is structured as follows: In the next section a brief quantitative assessment of the input and output of Japanese corporate R&D, particularly their recent development, is conducted. Thereafter, in the third section the managerial reorganization of the R&D in Japanese manufacturing is analyzed in qualitative terms, concentrating on the period since the second half of the 1980s. In the final section the results are summarized, and future implications for science and technology policy are pointed out.

Quantitative analysis: Japanese industrial R&D in the 1990s

In this section, a quantitative overview of the level and the development of R&D activities of Japanese manufacturing firms in the 1990s shall be given. The main focus of this assessment is on input oriented data compiled from statistical surveys. The results are complemented by the analysis of some output oriented indicators.

The main yardsticks for R&D input are R&D expenditure and R&D personnel. The overall amount of R&D expenditures of Japanese industrial firms in the 1990s is between 8 and 9 trillion Yen, the R&D intensity fluctuating around 3.5 percent of sales (Figures 8.1 and 8.2). This means that the proportion of sales dedicated to R&D is on a similarly high level as in other

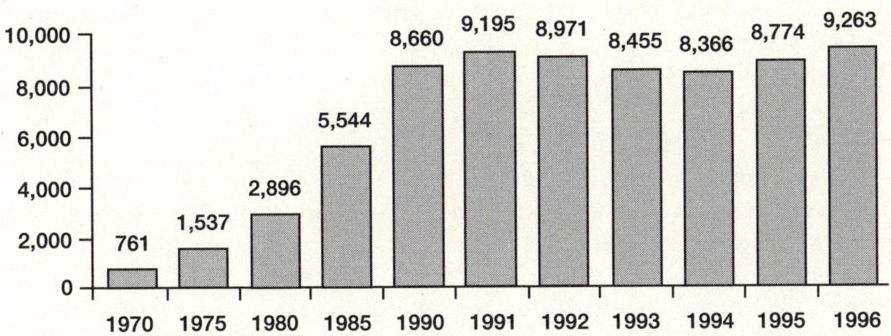

Figure 8.1 R&D expenditure of Japanese manufacturing firms, 1970–96 (billion Yen)
Source: Somucho Tokeikyoku (1997 and various editions).

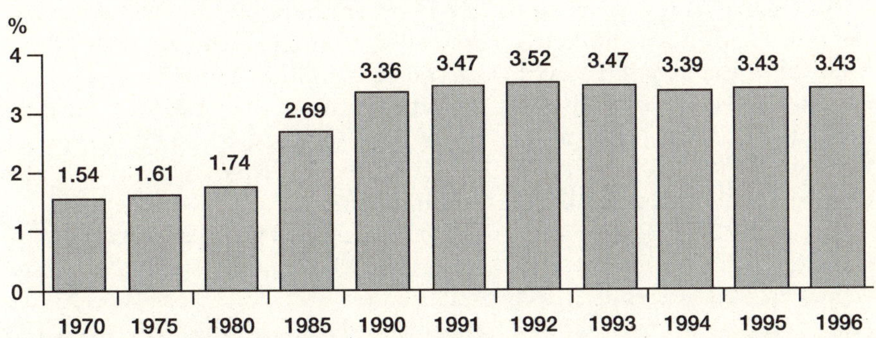

Figure 8.2 R&D intensity of Japanese manufacturing firms, 1970–96 (percent)
Source: Somucho Tokeikyoku (1997 and various editions).

leading industrialized countries.[1] From a quantitative viewpoint, there is no evidence that Japanese firms still rely on technology transfer from abroad, as may be indicated by a significantly lower R&D intensity than in the US and Western Europe.

Another interesting observation is the fact that in the 1990s a secular trend of steep expansion in Japanese industrial R&D has clearly come to an end. Between 1970 and 1990, R&D expenditure almost doubled every five years. The R&D intensity of sales also rapidly increased, particularly throughout the 1980s. After peaking in 1991, respectively in 1992, however, both indicators showed a slow decrease for several years. In 1995–6, absolute R&D expenditures rebounded, while R&D intensity remained at a lower level than at its previous peak.

One obvious explanation for the recent stagnation of industrial R&D efforts is the recession in the years after 1991. Since firms are under strong cost reduction pressure when demand is shrinking, a reduction in R&D expenditure is a normal phenomenon under such circumstances. Yet it should be noted that in earlier recessions, industrial R&D expenditures in Japan continued to increase. The fact that R&D expenditure has been shrinking for the first time since decades during a recession may be regarded as an indication that the previous period of catch-up has come to an end. Hence, it can be argued, the strategic behavior of Japanese firms is now approaching that of their Western counterparts.

What about the structure of industrial R&D expenditure? An indicator frequently used in this context is the distribution of R&D expenditure among basic research, applied research, and development. In the case of Japanese firms, the proportion of these three categories of R&D has remained almost unchanged since 1987, with around 6.5 percent of expenditure spent for basic research, around 22 percent for applied research, and the remainder for development (Somucho Tokeikyoku 1997: 18). Thus, concerning this indicator, there is no statistical evidence for recent structural change in Japanese corporate R&D.

When analyzing the industrial R&D expenditure by firm size, however, there is a remarkable shift toward smaller firms in recent years. While it declined in the large firms with a capital of 10 billion Yen or more by 11.2 percent between 1991 and 1994, the expenditure of the remaining, smaller, firms decreased only 2.5 percent in the same period.[2] Although industrial R&D in Japan still appears to be dominated by large corporations, the relative importance of small and medium firms is gradually increasing.

Regarding the development of R&D personnel in Japanese industrial firms, there has been a rapid expansion throughout the 1970s and 1980s as well as in R&D expenditures. In R&D personnel, however, the capacity for expansion has even persisted throughout the 1990s, with the exception of the year 1997 (Figure 8.3). This implies a major shift in the structure of industrial R&D by type of expenditure: whereas the proportion of personnel

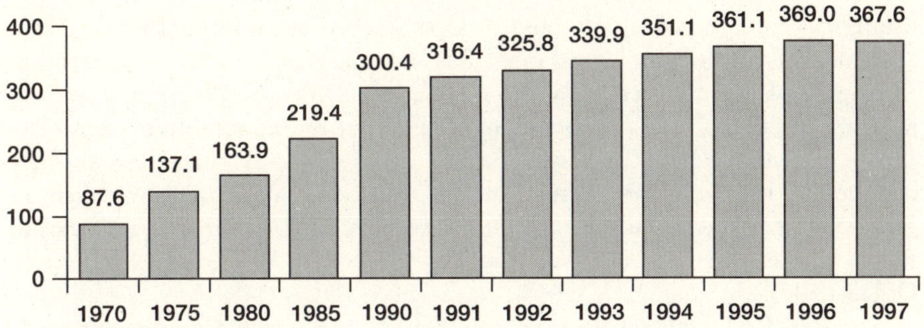

Figure 8.3 Number of researchers in Japanese manufacturing firms, 1970–97 (thousands)
Source: Somucho Tokeikyoku (1997 and various editions).

expenditure increased from 38.7 percent in 1991 to 45.2 percent in 1995, expenditure for capital investment was cut by more than one third in the same period, resulting in a decrease of overall share of expenditure from 15.5 percent to 9.3 percent (Somucho Tokeikyoku 1997: 82–4). In other words: the decline of R&D expenditure of Japanese industry in the 1990s is mainly due to a sharp decrease in capital investment, whereas the R&D personnel still continued to increase. The recruitment of new R&D personnel however has sharply decreased in recent years (Kagaku Gijutsucho 1995: 8). The only remaining explanation for the overall increase in the R&D workforce is that personnel from other departments inside the firms have been moved to R&D units. This is a rational countermeasure during a recession under the premise that personnel cannot be reduced in the short term; unproductive workers from other departments may have been shifted temporarily to R&D as a means of vocational training. This kind of measure however, does not necessarily imply a strategic upgrading of R&D activities within the firms.

In addition to the input-oriented analysis conducted above, the level and development of the R&D output generated by Japanese manufacturing firms in the 1990s shall now briefly be taken into consideration.[3] An often used output indicator in the field of R&D is the number of patents that were applied for and registered. The absolute number of domestic patent applications is in Japan, due to some specifics of the Japanese patent law, extremely high by international standards. The development over time shows, however, that their quantitative level has remained very stable in recent years. The number of domestic registrations, as well as the amount of applications and registrations in foreign countries, is second in the world ranking between the US and Germany.[4] This is in accordance with the economic ranking of Japan's industry as well as with its R&D input level. In the 1990s, there is a slight decline in the overall number of patent registrations granted to Japanese.

In order to measure Japanese patent output compared to other countries without having to consider the peculiarities of the Japanese patent law, the number of foreign patents granted to Japanese can be additionally taken into account. As can be seen from Figure 8.4, the share of US patents granted to Japanese was rapidly increasing throughout the 1980s, resulting in a leading position of Japan in the proportion of US patents among foreign nations. The Japanese patent share however, peaked in 1993 and thereafter slightly declined. This indicates some loss in *relative* technological strength of Japanese firms, whereas the *absolute* numbers of US patents granted to Japanese was still increasing in the mid-1990s (US Patent and Trademark Office 1997).

Another R&D output indicator is the relation between the payments received from technology exports and the amount paid for technology imports. Japan was running considerable deficits in the so-called technology balance of trade until the 1980s, indicating that the technology transfer from Western countries played a significant role in the Japanese innovation system in the postwar decades. In the 1990s, Japan still has a significant technology import surplus with the US. On the other hand, technology exports to developing countries, mainly in Asia, have rapidly increased in recent years. Moreover, the technology trade with western European countries has become fairly balanced due to increased technology exports from Japan (Somucho Tokeikyoku 1997: 27–8). As a result, Japan's technology balance of trade as a whole has changed in the 1990s from a deficit to a surplus that is widening year by year.

A final indicator often used for R&D output performance is international market share in high-technology products. According to the OECD, the market share of Japanese companies in six product groups defined as technology intensive rose to over 25 percent in the whole OECD in the second half of the 1980s, positioning Japan second again between the US and Germany. In the 1990s, the position of Japanese industrial firms on the global high-tech markets remains strong (National Science Board 1993: 440–5; BMBF 1996b: 40).

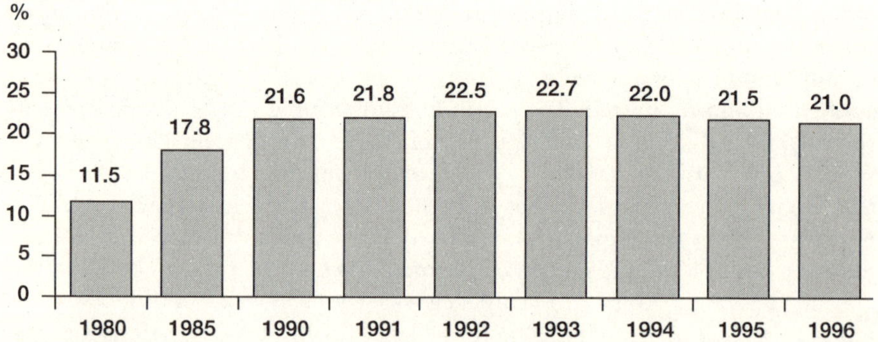

Figure 8.4 Share of US utility patents granted to Japanese (percent)
Source: Compiled from US Patent and Trademark Office (1997).

The quantitative analysis of the R&D input and output of Japanese manufacturing firms can be summarized as follows: First, the R&D input of Japanese industry is stagnating in the 1990s after rapid expansion in the preceding decades. Second, the R&D output remained fairly stable in the same period, although there are some differences in performance, depending on the indicator chosen for analysis. Third, the levels of input and output are by and large corresponding to each other, indicating that the efficiency of R&D in Japanese manufacturing firms is, when measured on a large scale, on a level that is comparable to their Western counterparts.

What results does the statistical analysis bear concerning firms' responses to the two main challenges identified above, strategic reorientation and cost reduction? On an aggregate level, the latter has been given clear priority. The firms reduced their R&D expenditure over several years, while maintaining their R&D output level at least in the short and medium term. There is little evidence, on the other hand, for a structural transformation of Japanese corporate R&D in the 1990s. The increasing relative importance of small and medium firms might be interpreted as a hint for such a reorientation, as small companies are generally regarded as more capable to perform flexible, innovative R&D than larger ones. The distribution between basic research, applied research, and development in industrial R&D, on the other hand, has remained largely unchanged in recent years.

The potential to analyze the strategies of manufacturing firms by using aggregated statistics is, however, fundamentally limited by the lack of accuracy of the data and the inability to grasp the actual processes on the micro-level that cause the changes of quantitative indicators. Therefore, in order to complement the statistical analysis, the managerial strategies Japanese manufacturing firms are recently pursuing in the field of R&D will be examined in the next section.

R&D reorganization strategies of Japanese manufacturing firms in the 1990s

The starting point: Japanese R&D management in the 1970s and 1980s

Before analyzing the transformation of Japanese industrial R&D in the 1990s, the base at which the companies began shall be pointed out shortly. A number of R&D management techniques and characteristics often described as typical Japanese have developed over the postwar decades and were refined during the 1970s and 1980s. They include:[5]

- a multi-layered, decentralized R&D organization structure in large firms;
- close cooperation between R&D units and other corporate departments;
- flexible allocation of personnel and budgets;

- phase-overlapping product development;
- a focus on incremental improvements (*kaizen* orientation);
- primarily vertical (and not horizontal) external R&D linkages; and
- a mainly domestic R&D system.[6]

Notably, it is widely assumed that these features *generally* apply to Japanese manufacturing firms. Although this can be disputed in several cases, the concept of a Japanese R&D management by itself implies that certain managerial commonalities among Japanese firms, in contrast to Western firms, do exist. There evolved a conventional wisdom throughout the 1970s and 1980s that the management of Japanese manufacturing firms, in the field of R&D as well as in general, is marked by a high degree of homogenity.

Despite their low degree of internationalization – a feature that is basically not regarded as favorable for R&D efficiency – Japanese firms achieved great technological success in the long term with their highly market-oriented R&D and innovation system. They managed successfully the high speed development of products that were well suited to market needs. This kind of R&D management, however, also bears a number of potential weaknesses. One such drawback is low priority of basic research, another one the tendency to over-compromise, since many departments and individuals are involved in the development process from an early stage.

The crucial question is: does the R&D management practiced by Japanese firms in the preceding decades still fit to the changed business conditions in the 1990s? As pointed out above, the end of the period of technological catch-up suggests that a reallocation of R&D efforts toward the strengthening of the domestic base of technology generation is needed. On the other hand, sluggish demand conditions and increased competition are pressing the firms for cost reduction and short and medium-term improvement of their profitability.

In the following sections, the recent transformation of Japanese industrial R&D will be examined in this context. The analysis focuses on three areas that are identified as primarily important to R&D management: organizational restructuring, R&D personnel management and internationalization of R&D activities. It is based on a number of recently disclosed questionnaire and case studies as well as on information directly gained from large firms' R&D managers.[7]

Transformation of the organizational structure

At the end of 1991, a comprehensive survey about strategic R&D management at large Japanese manufacturing companies was conducted. Since the majority of industrial R&D in Japan is performed by such large firms, the results of this questionnaire can be regarded as highly representative in this field. A main result was that about two thirds of the responding companies had a R&D department that was directly responsible to the president (NISTEP 1993: 11). This indicates that the function of R&D is regarded as an important one in most of the large Japanese firms. But how are the R&D activities in the firms actually organized?

The organizational structure of R&D in large Japanese manufacturing firms is often two or even three-layered: there are central R&D units, development centers attached to the strategic business units, and engineering units at the factory level (Akashi 1995: 13–14). This decentralized structure evolved throughout the preceding decades. The R&D units on the business line and factory level existed already in the postwar decades as supporting activities in the technological catch-up process. Since the 1960s, however, the need for a build-up of more basic-oriented, long-term R&D was acknowledged in a growing number of firms. An additional reason for this was the small extent of basic research being performed in publicly financed institutions in Japan. As a consequence, a large number of central company research laboratories was founded, particularly throughout the 1980s (Akashi 1995: 8–9; Sakakibara 1995: 164). Hence the organizational structure of R&D in many industrial firms was characterized by rapid growth and the build up of additional R&D units, while the companies as a whole were growing quickly. It has been reported, however, that in many cases the corporate R&D units have to a considerable extent been kept busy with conventional development activities passed over from the R&D units at the business line and factory level. Their real activities have not always been as centered on long-term projects as initially planned (Akashi 1995: 13–14).

How then was the reaction to the end of rapid growth and the recession after 1991? Several surveys show that in a large number of companies drastic R&D reorganization measures were taken. In a questionnaire sent by the Science and Technology Agency to large companies in 1994, 38.6 percent of the respondents stated that they had reorganized their R&D activities on a large scale in the last two to three years. Most of the remaining companies saw the necessity to do so in the near future (Kagaku Gijutsucho 1995: 14–15). Another survey conducted at year-end 1993 that covered large as well as medium-sized firms in five industries (steel, chemicals, machinery, electronics, and vehicles) led to similar results: 41 percent of the responding companies had recently reorganized their R&D, another 26 percent were considering such a reorganization (KSKKK 1994a: 22).

In the same survey, firms were also questioned about the content of the R&D reorganization measures they were carrying out or considering (Table 8.1). The definition of 'reorganization' was rather broad and therefore also covered actions that will be discussed in the following paragraphs. Concerning changes in the organizational structure, however, by far the most frequent answer was 'concentration and fusion of R&D units,' whereas less than one out of eight of the companies was conducting or considering the establishment of new R&D units. Moreover, considering the fact that close

Table 8.1 R&D reorganization measures carried out or considered by Japanese manufacturing firms, year end 1993

Strengthening of communication between R&D and business units	43.7%
Concentration and fusion of R&D units	29.9%
Promotion of mobility of researchers; flattening of R&D organizational structure	23.7%
Establishment of new R&D units	12.3%
Introduction of flexible working hours (*sairyō rōdōsei*)	9.0%
Downsizing of R&D units	4.8%
Introduction of annual salary system (*nenpōsei*)	4.8%
Internationalization	3.4%
Other	3.4%

Source: Compiled from KSKKK (1994a: 84).
Note: Because of multiple answers, the sum of all responses exceeds 100%.

cooperation between R&D and other departments is generally regarded as a feature of Japanese company R&D, it is quite surprising that the most frequent answer concerning R&D reorganization contained the further strengthening of inter-department communication.

Additional information can be gained from reports about the R&D reorganization of single large firms. As can be seen from Table 8.2, almost all of the large Japanese electronic companies have conducted reorganization measures in R&D throughout the 1990s. Concerning the strategic direction of the reorganization process, there are large discrepancies among the firms. However, the majority of them are apparently trying to increase the efficiency of their R&D activities by consolidating organizational units. The establishment of new, additional R&D units has become the exceptional case.

Another interesting point lies in the direction in which the R&D resources are reallocated. Again, the orientation of the firms are not all similar. But as a general trend there appears to be a shift from the corporate R&D centers to the 'traditional' development and engineering departments attached to the business units of the companies. This applies also for the firms interviewed by the author. With one exception, all of them conducted a major R&D restructuring aiming at the decentralization of related activities throughout the first half of the 1990s.

As a whole, the trend of the 1980s, when a large number of corporate R&D centers were founded, is now clearly reversed. Under the constrained business conditions of the 1990s, most of the firms concentrate their R&D activities on fewer units, mainly by the fusion of existing ones. Moreover, the priorities are laid again on the R&D activities that are directly connected to the business activities of the firms, whereas the resources devoted to corporate R&D centers are diminished. This is in accordance with the priorities expressed by many companies in the questionnaire survey for a stronger cooperation between R&D and other departments.

Table 8.2 R&D reorganization measures of Japanese electronic firms, 1992–96

Hitachi	Amalgamation of 2 corporate R&D centers; establishment of new development centers in each business unit (8/1993) Shift of 30% of the personnel working at corporate R&D labs to development centers at the business units (4/1995)
Toshiba	Reorganization of 12 research institutes into 6 new units: fusion of 2 central research institutes into 1 R&D center (10/1992) Establishment of 3 new R&D units in the microelectronics division; shift of 200 R&D workers from the corporate laboratory to division level units (4/1996)
Matsushita Electric	Reorganization of central research institute into technology support units for each business line (11/1993) Reorganization of central technology department into a basic research and an applied research department; fusion of 27 research institutes into 9 new institutes (2/1994)
NEC	Establishment of 7 new development institutes and 1 technology institute that are attached to the development and the business groups; relocation of R&D workers from the central R&D institute to the new units (8/1992)
Fujitsu	Reorganization of 3 regional and 5 functional research institutes into 4 new project units (6/1993)
Sony	Establishment of a new central development center (1/1993) Establishment of a R&D planning office to coordinate R&D activities on the business line level (12/1993)
Mitsubishi Electric	Reorganization of 4 central research institutes into a business operation unit; focus on product development (6/1993)
Sharp	Reorganization of 3 research institutes and 2 development centers at business units into 1 technology research institute and 4 product development institutes (4/1993)
Oki Electric	Establishment of a multimedia business promotion unit that is jointly staffed with business planning, engineering and research personnel (3/1994)
Sanyo Electric	Reorganization of 7 central research institutes into 4 new units; upgrading of 7 research centers and 1 technology center at business units to research institutes (12/1993)

Sources: KSKKK (1994b); Nikkei database; interviews with firms' R&D managers.

Apparently many Japanese companies, faced with strong pressures for cost reduction in the 1990s, have in terms of organizational structure returned to the kind of R&D management they exercised in the preceding decades. The proper functioning of development activities in cooperation between R&D and other company units has been given priority rather than the strengthening of more long-term based R&D activities in corporate R&D centers. Considering the earlier observation that the resources in the corporate R&D units had often been used for conventional development previously, it may be concluded that in a number of firms the organizational structure has just been readjusted to the real R&D activities being carried out.

Transformation of R&D personnel management

There are a number of features that are commonly associated with personnel management in large Japanese companies: lifetime employment, seniority-based compensation systems, and high inter-department mobility of workers within the firms. Concerning the field of R&D, empirical surveys from the 1980s suggest that the first two characteristics mainly apply: whereas a substantial number of workers in North-American and European firms' R&D departments had already changed their employer at least once, the corresponding figures for Japanese R&D workers were extremely low (Ernst and Wiesner 1994: 104; Sakakibara 1995: 23; Sato 1995: 46). The seniority-based compensation system generally applied by large Japanese firms covered the R&D departments as well (Ernst and Wiesner 1994: 146). The rotation of personnel between departments however, is in Japanese companies not necessarily higher than in their counterparts in other advanced countries. According to a comparative study conducted at the end of the 1980s, the percentage of researchers and developers with working experience in other departments was in US companies actually higher than in Japanese ones (Shapira 1995: 123–5).[8]

Consequently, the lack of continuity of work that is sometimes pointed out as a drawback of Japanese company R&D, may not really be more of a problem than in any other industrial country. The results of empirical surveys rather suggest that in many Japanese firms, the individual treatment of R&D workers may cause incentive problems: many Japanese company researchers expressed frustration about excessive working hours, too little freedom of work, low salaries, and poor working equipment (Ernst and Wiesner 1994: 150; Shapira 1995: 137). The R&D workers in Japanese manufacturing companies should mostly be able to specialize and acquire advanced technical skills as much as their colleagues in North America and Europe. Many of them however, seem to be discouraged because of lack of individual incentives at their working place. This causes the danger of low R&D productivity, particularly in basic-research oriented fields where individual performance is of predominant importance.

What kind of measures have been carried out by Japanese manufacturing firms in the 1990s in order to improve their R&D personnel management? One organizational strategy often mentioned in this context is the creation of individual career paths for technical specialists as an alternative to general managerial career paths. In the above-mentioned 1991 survey among large firms, more than half of the respondents stated that in their company specialized posts for senior R&D workers are provided. However, even in this case there remains the possibility that the treatment of R&D specialists is eventually quite similar to that of the general administrative staff, i.e. not preferential to less qualified employees (NISTEP 1993: 48–9).

On a more concrete level, other actions taken by Japanese companies to improve individual incentives for R&D workers can be observed. Mostly discussed are the introduction of flexible working hours (*sairyō rōdōsei*) and of annual, i.e. individual-based salary systems (*nenpōsei*). In the above-cited 1993 survey, close to one out of ten companies stated it introduced or was considering flexible working hours in the R&D department, while one out of twenty did so with individual-based annual salaries (Table 8.1). This clearly shows that the transformation of R&D personnel management had up to then only been the concern of a rather small minority of manufacturing firms.

More recent studies show that a number of large manufacturing firms have introduced flexible working hours and performance-based compensation systems for some of the employees in their R&D departments. These include Mitsubishi Electric, NEC, Fujitsu, Yokokawa Electric, Nippon Denso, and Matsushita (KSKKK 1994b: 88; Sakakibara 1995: 263). Some firms carried out additional measures in order to improve the attractiveness of their R&D. Fujitsu is offering three-year-sabbaticals for researchers in its corporate R&D laboratory who propose promising research projects. Moreover, the company is hiring parts of its R&D staff not – as conventionally done – in April, but throughout the whole year to attract people who already have working experience in technical fields. Sony is now engaging 10 percent of its new technical staff solely on the base of individual criteria – in contrast to the prior practice that gave recommendation letters submitted by universities a decisive role in the selection process (Sakakibara 1995: 262–3; Nihon Keizai Shinbun, 16 April 1996).

Overall, it can be said that some of the large Japanese manufacturing firms, particularly electronic firms, are seriously attempting to transform their R&D personnel management toward more individual-based incentive structures. But it also cannot be overlooked that these measures are still carried out only by a minority of companies, and even in these firms they are mostly applied to some sections of the R&D personnel only. According to R&D managers interviewed by the author, a major obstacle appears to be the general system of personnel management practices in large Japanese firms that is still largely based on seniority and long-term employment. It proves difficult to create special incentive structures only for the R&D staff while keeping the old system for all other employees in different departments. Altogether, the transformation of R&D personnel management in manufacturing firms has to be regarded as a process that is still in its infancy.

Internationalization of industrial R&D

A final issue to be analyzed is the internationalization of R&D activities by Japanese manufacturing firms. The internationalization strategy is somewhat fundamentally different from the measures discussed above. Those transformation strategies are primarily aimed to increase the productivity and efficiency of the R&D performed in Japan. Therefore, they can be regarded as efforts to overcome drawbacks in the domestic R&D system. Internationalization of R&D means in contrast creating a strategy to circumvent rather than to overcome these weaknesses by using comparative advantages for R&D at foreign locations.

As already mentioned, the internationalization of Japanese industrial R&D has been traditionally very low, particularly when compared with European countries (Belderbos 1995: 4). Recently however, there are growing concerns in Japan that the 'hollowing out' process in the manufacturing sector may have spread to the field of R&D. The questions to be answered are to what extent Japanese manufacturing firms are really shifting R&D activities to locations abroad, and what functions these foreign R&D sites primarily have.

First of all, it should be noted that the term 'internationalization' of R&D comprises several meanings. One strategy adopted by a number of Japanese firms is to fund foreign university research facilities in order to profit from the results generated there. A well-known case is MIT where nineteen out of fifty-five corporate-endowed chairs are from Japanese companies (Herbig 1995: 222). Another means to gain access to technological knowledge in other countries are strategic alliances with foreign companies. Between US and Japanese firms alone, hundreds of such alliances exist, indicating their relative importance (Osborn and Baughn 1995: 94–9).

The analysis in this paragraph, however, focuses on the number of foreign R&D units owned by Japanese companies and the amount of R&D expenditures spent abroad. An official source concerning these indicators are the foreign investment statistics held by the Ministry of International Trade and Industry (MITI). According to these statistics, the foreign affiliates of Japanese manufacturing firms spent approximately 150 billion Yen in R&D in 1994, more than 90 percent of it in Europe and North America (Figure 8.5). The R&D funds spent abroad amounted to 1.8 percent of the domestic R&D expenditure of the companies surveyed, up from 0.8 percent in 1986 (TSSS 1996: 60). This figure appears not to be very high and is, indeed, much lower than in the case of manufacturing firms in Western countries. US firms, for instance, spent 10 percent of their R&D in foreign countries in the 1990s (National Science Board 1996: 161), the corresponding figure for German firms is 15 percent (BMBF 1997: 54). Compared with eleven years earlier however, the amount of Japanese industrial R&D spent abroad grew more than tenfold. The number of foreign research sites increased dramatically

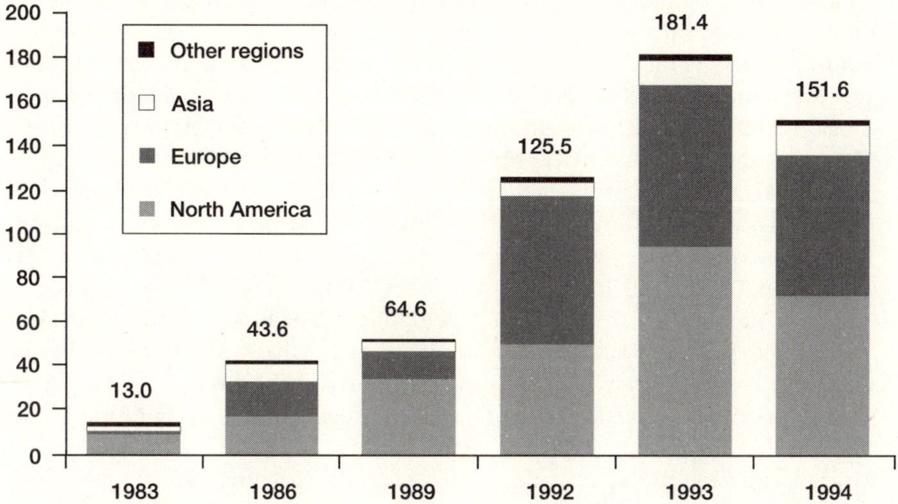

Figure 8.5 R&D expenditures of foreign affiliates of Japanese manufacturing firms (billion Yen)
Sources: TSSS (1994 and various editions, 1996 and various editions).

from 88 in 1987 to 236 in 1993 (KSKKK 1995: 60–4). Overall, the statistics indicate a rapid increase in the relative importance of foreign R&D for Japanese manufacturing firms in recent years.[9] This share is, however, still much smaller in Japan than in other advanced countries.

In the MITI foreign direct investment statistics, the firms are also asked about the functions of their overseas R&D facilities. The most frequent responses in the recent surveys were the 'support of local production activities' and 'product development for local markets,' followed by 'support of local marketing activities' and 'collection and analysis of information about the local site.' The 'development of products for the global market' and 'basic research' on the other hand, accounted for only 6.5 percent of total responses (TSSS 1994: 36). These results suggest that most of the R&D activities of Japanese firms in foreign countries are closely connected to the internationalization of their production and marketing activities. A truly global innovation strategy that may be indicated by the usage of comparative advantages of R&D sites in different countries for basic research and global product development still appears to be limited to relatively few cases.

Since the R&D investments by Japanese manufacturing firms at foreign sites are highly dynamic, however, the additional consideration of recent data is valuable in order to obtain an accurate picture of the current state of affairs in this field. In the questionnaire survey among medium and large firms from year-end 1993 mentioned above, 30.3 percent of the respondents stated that they maintain R&D activities in foreign countries. The most important motives were, again, the product development for local markets, the collection of relevant information at the local site, and the support of local production facilities. Some 18 percent of the companies engaged in R&D abroad, however, also stated that they conduct genuine research at their foreign locations (KSKKK 1994a: 86). An interview survey from 1995 among seventeen electronic and pharmaceutical firms with fifty-six foreign R&D sites draws a rather heterogeneous picture (Kiba 1996: 13–28). Concerning the size of the R&D units abroad, there was a rather even distribution from very small units with less than nine employees up to large ones with more than sixty employees. About 40 percent of the foreign sites were mainly engaged in research, the remainder mainly in development. The field of R&D most often mentioned was software development, an area where other industrialized countries are commonly believed to have a competitive advantage over Japan. The respondents were also asked to evaluate their foreign R&D units. While the results of development activities performed abroad were assessed as slightly inferior to Japan, in the area of research some superiority of the foreign sites was noticed.

There also exist case studies about several large manufacturing companies like Nissan, Canon, Sony, Kao, and NEC (Takahashi 1996: 79–134). The results suggest that these firms are quite advanced on their way toward global innovation strategies.

Altogether, recent studies confirm that the internationalization of Japanese industrial R&D is highly dynamic. An increasing number of firms are making considerable efforts in order to appropriate foreign resources for their R&D strategies. This result is plausible in the sense that it may be indeed easier for many firms to compensate for weaknesses in their domestic R&D by building up or acquiring research institutes abroad than by transforming their existing organization at home.

There might even exist some potential to speed up the reorganization of domestic R&D activities through internationalization. Sakakibara, Westney and Kosaka (1996: 70) regard it as possible to increase the productivity of domestic R&D by tearing down the organizational and processual isomorphism within Japanese firms. On the other hand, internationalization is also likely to cause new managerial problems for the firms. Odagiri and Yasuda (1997: 225–6) specifically address the fields of integrating the foreign units into the overall management system and of personnel management for the staff of the foreign R&D units. As discussed above, the management of Japanese corporate R&D relies to a large extent on close interfunctional linkages and on specific tools of personnel management. Both cannot be expected to be transferred easily to R&D outlets in Western countries.

These two views are addressing opposite sides of the same coin. Internationalization creates new opportunities for Japanese corporate R&D, but it exposes it at the same time to new managerial challenges. It appears therefore that this area has in the long term probably the most far-reaching implications among the managerial transformations discussed above. Nevertheless, it has to be kept in mind that the degree of internationalization of R&D activities is still much lower in Japanese firms than in their counterparts in the Western advanced countries. In this sense, the Japanese firms are latecomers once again, as well as with the internationalization of their production facilities that began some years earlier.

Conclusions and implications

The analysis of the transformation of Japanese industrial corporate R&D performed above led to quite mixed results. A quantitative assessment of the R&D input and output of Japanese industry in the 1990s showed that while R&D expenditure decreased somewhat in the years after the burst of the bubble economy, they nevertheless remained at a high level showing Japanese companies to be in the first row of industrialized countries. The R&D output also remained largely stable.

The qualitative study of the transformation of Japanese company R&D management showed, on the other hand, that the strategic direction of the reorganization measures is quite heterogeneous. Concerning *organizational structure*, there appears to be a shift of resources away from corporate R&D centers back to R&D units at the business line and factory level. Hence, in this field no focus on the strengthening of long-term research is noticeable. The measures may be better interpreted as an adjustment of the organizational structure to the R&D activities actually performed that was instigated by increasing cost and profitability pressure in the years after the bubble.

In the field of *R&D personnel management*, some large firms are recently moving toward more individual-based, flexible working conditions and compensation systems for parts of their R&D workforce. In the long term this may contribute to an increase of motivation, creativity, and productivity in basic-oriented research. However, these measures are still limited to a comparatively small number of companies, and even in these firms often apply only to parts of the R&D workforce. Therefore, their impact should not be overestimated for the time being.

The *internationalization of industrial corporate R&D* is, on the other hand, in dynamic progress, although from a very low starting point. R&D expenditure and R&D sites owned by Japanese companies in foreign countries are rapidly increasing. Most of these activities are directly connected with production and marketing bases at foreign locations, but the number of institutes primarily designed for research is also rapidly expanding. Overall, it appears that the field of internationalization of R&D is by far the most dynamic among the areas that were analyzed.

Summarizing the results, it can be concluded that the Japanese manufacturing industry is responding with a combined R&D reorganization strategy to the two main challenges it is facing in the 1990s. R&D cost reduction efforts and the shift of resources from corporate R&D centers to units directly connected to product development indicate a strategic redirection on what may be regarded as the managerial core competence of Japanese industrial R&D: the rapid development of products that are well-suited to market needs. Simultaneously, at least some of the firms are attempting to transform their R&D personnel management toward more individual- and performance-based incentives for some of their R&D workers in order to improve their long-term productivity. In addition, an increasing number of firms are pursuing the internationalization of R&D activities as a means to circumvent structural shortcomings in the Japanese institutional setting for R&D, and possibly even to reform their overall R&D system in the long run.

From the viewpoint of Western R&D managers, it may be concluded that on the whole, the profile of strengths and weaknesses of their Japanese competitors will probably not change fundamentally in the field of R&D for the time being. Although there have been some significant adjustments made in the management of domestic R&D, there is little evidence for anything like a structural overhaul of the Japanese corporate R&D management system. Moreover, certain developments are not necessarily specific to Japanese firms. For instance, there seems to be a world-wide trend among large manufacturing firms to streamline their central R&D laboratories (BMBF 1997: 43). The only development that can be expected to bear far-reaching implications for Japanese corporate R&D is its increasing internationalization. As described above, however, most Japanese firms are still lagging behind their Western counterparts in this field.

This does only apply on the whole, however. Another fact to be noted in this context is the increasing diversity of managerial strategies among Japanese manufacturing firms. Although certain overall tendencies do exist concerning the strategic direction of R&D management in Japanese manufacturing, there are surprisingly strong differences between individual firms. Among the firms surveyed by the author, some strongly emphasized the internationalization strategy, while others focused on the reorganization of their domestic R&D activities, particularly concerning personnel management practices. Similarly, while some of them made considerable cuts into their R&D budget in the first half of the 1990s, others maintained the level of their R&D expenses throughout the whole recession. Assumed that these differences in the strategic orientation of firms prevail throughout the next years, an increasing variety may also be witnessed concerning the performance of Japanese manufacturing firms at the beginning of the twenty-first century. Some of them may become truly global firms, as some of their Western counterparts, while others may remain firmly rooted in their home country. In other words: the conventional wisdom that most of the Japanese manufacturing firms are highly similar concerning their management system may, at least in the field of R&D, have to be revised.

Finally, there are two implications for Japanese science and technology policy. First, the potential for Japanese manufacturing firms to transform their domestic R&D management appears to be limited. In particular, there are natural boundaries for further increases in basic research. After all, private firms are under continuous cost and profitability pressure. Since the output of basic research is partially a public good, their potential to perform such activities is fundamentally restricted. Also it should be kept in mind that in Japan the proportion of basic research performed by the private sector is already significantly higher than in other industrialized countries (Kagaku Gijutsucho 1996: 393–7). Hence, if there is a lack of basic research in Japan, this gap has primarily to be filled by publicly financed institutions and not by private firms.

Second, the dynamics and the by and large favorable evaluation of the international R&D activities of these companies show that the international division of labor may be a viable means to overcome certain problems resulting from institutional shortcomings on the national level. Through the good performance of some foreign R&D sites, not only the companies who undertake these activities, but also the whole country should benefit in the longer term. Therefore, rather than becoming too anxious about the 'hollowing out' of R&D, firms who are actively engaging in international R&D may be encouraged to do so.

NOTES

1 Refer for R&D intensity of sales in the German industry to BMBF (1996a: 562–3) and the US industry to National Science Board (1996: Appendix: 119–20).
2 Compiled from Somucho Tokeikyoku (1997 and various editions).
3 The output measurement in the field of R&D is in general very problematic. The specific shortcomings of the various indicators are not discussed here. Refer for a general assessment of the output measurement in R&D Patel and Pavitt (1987: 6–9).
4 Detailed statistics of these indicators can be found in Nihonkoku Tokkyocho (1997) and NISTEP (1997: 182–93).
5 Refer for an overview to Itami (1987: 95–109); Odagiri and Goto (1993: 105–9); Wakasugi (1994: 165–71); Nonaka (1994); Westney (1994: 166–73); Okimoto and Nishi (1994: 186–206); Nishiguchi and Ikeda (1996); Laage-Hellman (1997: 133–49).
6 It goes without saying that in the postwar technology catch-up process, many Japanese firms relied heavily on foreign technology. However, in most cases actual R&D activities conducted by the firms were almost exclusively performed in-house, or, in the case of R&D cooperation with other institutions, were mostly Japanese.
7 Between summer 1996 and spring 1997, the author conducted explorative interviews about R&D restructuring at five medium-sized and large Japanese firms in the machinery, electronic, and automobile industries.
8 According to the survey analysis of Ernst and Wiesner (1994: 125), rotation of R&D personnel is in Japanese companies somewhat higher than in US and German companies. Nevertheless, their analysis indicates as well that there is no

fundamental difference in the frequency of R&D personnel rotation between the three countries. In a recent case study about personnel management practices of a large Japanese firm, Kusunoki and Numagami (1997) stress the importance of interfunctional transfer of engineers for R&D management. Their research shows at the same time, however, that these transfers mostly occur throughout the first years after entering the firm (Kusunoki and Numagami 1997: 182–3). This suggests that they are largely a part of the general training programs for newly hired staff that is common in large Japanese firms.

9 The amount of R&D performed abroad may even be significantly underestimated in the MITI surveys. For about 75 percent of the foreign affiliates, the question about R&D spending remains unanswered. This does not necessarily mean that in these affiliates no R&D is performed, as assumed in the official statistics. To review these measurement problems, refer to Belderbos (1995: 6–8).

REFERENCES

Akashi, Y. (1995) Nihon kigyō no kenkyū kaihatsu, gijutsu kaihatsu, seihin kaihatsu (R&D, technology development, and product development in Japanese enterprises). In Akashi, Y. and H. Ueda (eds), *Nihon kigyō no kenkyū kaihatsu shisutemu* (The R&D system in Japanese enterprises). Tokyo: Tokyo Daigaku Shuppankai, pp. 3–30.

Belderbos, R. (1995) *Overseas R&D by Japanese Electronics Multinationals. Evidence From the Basic Survey 1992*. Mimeo, Institute for Economic Research, Hitotsubashi University.

BMBF – Bundesministerium für Bildung, Wissenschaft, Forschung und Technologie (ed.) (1996a) *Bundesbericht Forschung 1996*. Bonn: Bundesministerium für Bildung, Wissenschaft, Forschung und Technologie.

—— (1996b) *Zur technologischen Leistungsfähigkeit Deutschlands*. Bonn: Bundesministerium für Bildung, Wissenschaft, Forschung und Technologie.

—— (1997) *Zur technologischen Leistungsfähigkeit Deutschlands. Aktualisierung und Erweiterung 1996*. Bonn: Bundesministerium für Bildung, Wissenschaft, Forschung und Technologie.

Ernst, A. and G. Wiesner (1994) *Japans technische Intelligenz. Personalstrukturen und Personalmanagement in Forschung und Entwicklung*. Munich: ifo Institut für Wirtschaftsforschung (= ifo studien zur japanforschung 7).

Francks, P. (1992) *Japanese Economic Development. Theory and Practice*. London and New York: Routledge.

Herbig, P. (1995) *Innovation Japanese Style. A Cultural and Historical Perspective*. Westport and London: Quorum.

Itami, H. (1987), *Mobilizing Invisible Assets*. Cambridge and London: Harvard University Press.

Kagaku Gijutsucho (ed.) (1995) *Minkan kigyō no kenkyū katsudō ni kan suru chōsa hōkoku* (Report on the research activities of private enterprises). Tokyo: Okurasho Insatsukyoku.

—— (1996) *Heisei 8-nenpan kagaku gijutsu hakusho* (White Paper on Science and Technology 1996). Tokyo: Okurasho Insatsukyoku.

Kiba, T. (1996) *Nihon kigyō no kaigai ni okeru kenkyū kaihatsu no pafōmansu ni kan suru chōsa* (Survey of the performance of Japanese company R&D abroad). Tokyo: Kagaku Gijutsu Seisaku Kenkyusho (NISTEP Report no. 43).

KSKKK – Kikai Shinko Kyokai Keizai Kenkyusho (ed.) (1994a) *Kenkyū kaihatsu no keizaiteki kōka ni kan suru chōsa* (Investigation of the economic effects of R&D). Tokyo: Kikai Shinko Kyokai Keizai Kenkyusho (unpublished).

—— (1994b): *Minkan kigyō no kenkyū kaihatsu katsudō ni kan suru kiso chōsa* (Basic investigation of the R&D activities of private enterprises). Tokyo: Kikai Shinko Kyokai Keizai Kenkyusho (unpublished).

—— (1995) *Minkan kigyō ni okeru gijutsu no kūdōka ni kan suru chōsa kenkyū* (Report on the technological hollowing out of private enterprises). Tokyo: Kikai Shinko Kyokai Keizai Kenkyusho (unpublished).

Kusunoki, K. and T. Numagami (1997) Intrafirm Transfers of Engineers in Japan. In Goto, A. and H. Odagiri (eds), *Innovation in Japan.* Oxford: Clarendon Press, pp. 173–203.

Laage-Hellman, J. (1997) *Business Networks in Japan. Supplier-customer Interaction in Product Development.* London and New York: Routledge.

National Science Board (ed.) (1993) *Science & Engineering Indicators 1993*. Washington, DC: US Government Printing Office.

—— (1996) *Science & Engineering Indicators 1996*. Washington, DC: US Government Printing Office.

Nihon Keizai Shinbun 16 April 1996, p. 1.

Nihonkoku Tokkyocho (ed.) (1997) *Tokkyochō nenpō. Dai 49-kan, Heisei 8-nenpan* (Annual Report of the Patent Office, 49 edition, 1996). Tokyo: Nihonkoku Tokkyocho.

Nishiguchi, T. and M. Ikeda (1996) Suppliers' Process Innovation: Understated Aspects of Japanese Industrial Sourcing. In Nishiguchi, T. (ed.), *Managing Product Development.* New York and Oxford: Oxford University Press, pp. 206–30.

NISTEP – Kagaku Gijutsu Seisaku Kenkyusho (ed.) (1993) *Nihon kigyō ni miru senryakuteki kenkyū kaihatsu manejimento* (Strategic R&D management at Japanese enterprises). Tokyo: Kagaku Gijutsu Seisaku Kenkyusho.

—— (1997) *Heisei 9-nenpan kagaku gijutsu shihyō* (Science and Technology Indicators, 1997 Edition). Tokyo: Kagaku Gijutsu Seisaku Kenkyusho (NISTEP Report no. 50).

Nonaka, I. (1994) Product Development and Innovation. In Imai, K. and R. Komiya (eds), *Business Enterprise in Japan.* Cambridge and London: MIT Press, pp. 209–21.

Odagiri, H. and A. Goto (1993) The Japanese System of Innovation: Past, Present, and Future. In Nelson, R.R. (ed.), *National Innovation Systems. A Comparative Analysis.* New York and London: Oxford University Press, pp. 76–114.

—— (1996) *Technology and Industrial Development in Japan. Building Capabilities by Learning, Innovation, and Public Policy.* Oxford: Clarendon Press.

Odagiri, H. and H. Yasuda (1997) Overseas R&D Activities of Japanese Firms. In Goto, A. and H. Odagiri (eds), *Innovation in Japan.* Oxford: Clarendon Press, pp. 204–28.

Okimoto, D.I. and Y. Nishi (1994) R&D Organization in Japanese and American Semiconductor Firms. In Aoki, M. and R. Dore (eds), *The Japanese Firm. Sources of Competitive Strength.* Oxford: Oxford University Press, pp. 178–208.

Osborn, R.N. and C.C. Baughn (1995) Governing United States – Japan High-Technology Alliances. In Liker, J.K., J.E. Ettlie, and J.C. Campbell (eds), *Engineered in Japan. Japanese Technology-Management Practices.* New York and Oxford: Oxford University Press, pp. 93–112.

Patel, P. and K. Pavitt (1987) Is Western Europe Losing the Technological Race? In Freeman, C. (ed.), *Output Measurement in Science and Technology. Essays in Honor of Yvan Fabian.* Amsterdam etc.: North-Holland, pp. 5–31.

Sakakibara, K. (1995) *Nihon kigyō no kenkyū kaihatsu manejimento* (R&D management in Japanese firms). Tokyo: Chikura Shobo.

Sakakibara, K., D.E. Westney and M. Kosaka (1996) International Product Development of Japanese Firms: Product Group Coherence and Internal Isomorphism Matrix. In Nishiguchi, T. (ed.), *Managing Product Development.* New York and Oxford: Oxford University Press, pp. 57–71.

Sato, H. (1995) Corporate Careers of R&D Personnel. In Shapira, P. (ed.), *The R&D Workers. Managing Innovation in Britain, Germany, Japan, and the United States.* Westport and London: Quorum, pp. 45–58.

Seisansei Kenkyusho (ed.) (1996) *R&D senryaku to jinzai kaihatsu. Nihon kigyō no aratana shinrō* (R&D strategy and personnel development. New orientations for Japanese firms). Tokyo: Shakai Keizai Seisansei Honbu.

Shapira, P. (1995) Comparisons and Contrasts in R&D Management in the United States and Japan. In Shapira, P. (ed.), *The R&D Workers. Managing Innovation in Britain, Germany, Japan, and the United States.* Westport and London: Quorum, pp. 121–42.

Somucho Tokeikyoku (ed.) (1997) *Heisei 8-nen kagaku gijutsu kenkyū chōsa hōkoku* (Report on the Survey of Research and Development 1996). Tokyo: Nihon Tokei Kyokai (published annually).

Takahashi, H. (1996) *Kenkyū kaihatsu kokusaika no jissai* (The reality of R&D internationalization). Tokyo: Chuo Keizaisha.

TSSS – Tsusho Sangyosho Sangyo Seisakukyoku (ed.) (1994) *Dai 5-kai kaigai tōshi tōkei sōran* (5th Statistical Overview of Foreign Direct Investment). Tokyo: Okurasho Insatsukyoku (published triannually).

—— (1996) *Dai 25-kai waga kuni kigyō no kaigai jigyō katsudō* (Foreign Business Activities of Japanese Enterprises. 25th edition). Tokyo: Okurasho Insatsukyoku (published annually).

US Patent and Trademark Office (ed.) (1997) *Patent Counts by Country/State and Year, Utility Patents, January 1, 1963 – December 31, 1996.* Washington, DC: US Patent and Trademark Office.

Wakasugi, R. (1994) Organizational Structure and Behavior in Research and Development. In Imai, K. and R. Komiya (eds), *Business Enterprise in Japan.* Cambridge and London: MIT Press, pp. 159–77.

Westney, D.E. (1994) The Evolution of Japan's Industrial Research and Development. In Aoki, M. and R. Dore (eds), *The Japanese Firm. Sources of Competitive Strength.* Oxford: Oxford University Press, pp. 154–77.

9

COLLABORATIVE RESEARCH IN JAPAN AND THE WEST

A case study of Britain's response to MITI's Fifth Generation Computer Initiative

Tim Ray

Introduction[1]

During the 1980s, growing awareness of Japan's 'economic miracle' stimulated Western interest in the alluring prospect of 'learning from Japan' and extracting the 'secret ingredients' of its success. In particular, paranoia about the ability of the Ministry of International Trade and Industry (MITI) to orchestrate technological change encouraged an enthusiasm for imitating Japanese government policy. One prominent example of this paranoia was MITI's capacity to cause shock waves of concern to reverberate around competitor nations through its announcement, in 1981, of a ten-year (1982–92) Fifth Generation Computer Systems (FGCS) initiative. This bold venture aimed to use cooperative research – conducted amongst competitor companies, working together in a specially formed 'Research Association' (RA)[2] – to produce 'thinking' computers based on artificial intelligence. It appeared that Japanese computer makers, having caught up with the West, were poised to become global leaders in a new era of information technology (IT). Although Western apprehension about the Fifth Generation threat receded long before the ten-year initiative had run its course, the immediate impact of MITI's original announcement was spectacular.

For many in the West, it seemed that a series of MITI RAs during the 1960s and 1970s had underpinned Japan's rapid progress in computer technology. In particular, the apparent success of MITI's 1976–80 VLSI (Very Large Scale Integration) RA encouraged a belief that Japan was somehow exploiting 'horizontal' cooperative research conducted amongst rival companies to fuel competitive advantage. Consequently, the FGCS project's planned use of cooperative research tended to give it added credibility.

Although MITI invited other countries to join its quest for Fifth Generation computing, one British academic suggested that this would be like: 'collaborating with a vacuum cleaner' (Oakley and Owen 1989: 17). Filled with consternation, Japan's competitors opted to 'fight fire with fire' and rushed headlong to organize countermeasures (Ferné 1989: 10). For example, in 1984, Europe launched the European Strategic Programme for Research in Information Technology (ESPRIT) while in the US, the formation of the Microelectronic and Computer Technology Corporation in 1982 – as a research center funded jointly by private firms – spearheaded a relaxation of the country's strict Anti-Trust Laws with the passage of the 1984 National Cooperative Research Act. This paved the way for further collaborative ventures. By 1989, 137 consortia had registered under the Act (Evan and Olk 1990: 37), including the 1987 foundation of 'SEMATECH' as a major Department of Defense (DoD) funded initiative to enable US firms to conduct joint research in the manufacture of semiconductor devices.

Using the example of the UK 1983–1988 Alvey Programme for Advanced Information Technology, which was in the vanguard of responses to Japan's 'challenge,' this chapter argues that it is, at best, difficult to develop British policy by imitating Japan. Historical differences between Britain and Japan have produced two very different national knowledge-creating systems. Moreover, these systems show signs of moving in different directions. After a long history of applied R&D – that served Japan well during its catch-up era – the country has been taking steps to complement a high level of industrial R&D with increased spending on more basic research conducted in the public sector. By contrast, Britain has long-standing strengths in basic research. After the Second World War, the significance of science-related technologies – such as nuclear power, radar, synthetic fibers and jet engines – encouraged a linear model of innovation in which scientific discoveries provided the seeds of economic growth. It was not until the mid-1960s that studies of technological change started to draw attention to the nature and significance of 'market pull' in the innovation process. Given the differences between the British and Japanese innovation systems, it is curious that the early-1980s found both countries embarking on large-scale 'pre-competitive' cooperative research initiatives in computer and related technologies. Even so, this could be more of a 'cross-over point' than evidence of convergence. Subsequently, Japan continued to expand its efforts in basic science, while the UK has sought to enhance links between public sector research and 'wealth creating processes.'

Arguably, the West's sudden enthusiasm for 'Japanese-style' cooperative research is surprising for at least three reasons. First, collaboration is potentially problematic: intertwined with the idea of legitimacy come difficulties arising from trust and concern about 'free-riders.' Such projects exist in the 'no man's land' between organizations, where cooperative activities lack the legitimacy of inside projects. Consequently, the effort of 'making the collaboration work' can all too easily tower above its achievements. Second, existing experience of collaboration in Europe often involved projects that were manageable but somewhat lackluster – and hardly appeared to be the basis for organizing major initiatives to promote industrial competitiveness. For example, a 1965 report on Europe's industrial research associations quoted one commentator who said that their business, though worthy, 'seems hardly more exciting than the work of the local Post Office' (Woodward 1965: 38). Third, a review of MITI's computer RAs before FGCS, reveals that – for the most part – they were neither pre-competitive nor collaborative. Participating firms tended to work on their own free-standing research modules, with MITI acting as an 'honest broker' that arranged the distribution of subsidies for mainly applied R&D. The FGCS project changed this by championing basic research and a unified approach to joint knowledge creation amongst rival companies. But Japan had no track record in this type of RA and participants were generally skeptical of a venture that threatened to compromise their independence and lacked a clear path to exploitation.

The present argument does not seek to detract from Alvey's undoubted achievements but rather to question the logic of basing British IT policy on MITI's computer RAs. Indeed, Alvey could be a case of 'doing the right thing for the wrong reasons.'[3] It helped to raise morale, provided valuable training opportunities, promoted common IT standards, propagated a so-called 'collaborative culture' (that helped firms to bid for pan-European cooperative initiatives), and fostered informal communication networks that are likely to be of lasting significance, especially as younger researchers rise to positions of greater importance. While these results might more than justify the effort and expenditure involved, Alvey's *raison d'être* was to enhance Britain's ability to compete in IT and – in this crucial respect – it generally failed to match expectations.

The UK delegation to MITI's 1981 Tokyo conference evidently concluded that one of the most useful lessons to be learned concerned Japan's organization of IT for the 1990s: 'In many important respects the Japanese approach to IT could be taken as a model for what any country would need to do if it was to be credible and competitive in IT over the next ten years or so' (Oakley and Owen 1989: 25). Given the nature and extent of differences between the Japanese system and Anglo-American capitalism, it seems strange that the delegation should envisage that Japan could provide a model for 'any country.' If the 'house of capitalism' has many rooms, the Japanese version perhaps lies in an annex that has always operated according to different rules and procedures. And an appreciation of the nature and implications of these differences is a vital precursor to drawing meaningful lessons from Japan.

This chapter's second section draws attention to differences between British and Japanese national innovation systems and the third section summarizes the evolution of attempts by MITI's computer RAs to catch up with and overtake the West. The fourth section reflects on Japan's increasing commitment to basic science, which is an important factor in understanding the broader policy context of MITI's move toward more basic research for computing. The fifth section reviews Britain's response to Japan's Fifth Generation challenge and British policy's subsequent emphasis on closer links between public sector research and 'wealth creation.' With the respective experiences of Japan and Britain in mind, the sixth section considers briefly the problems of using pre-competitive research cooperation as a vehicle for promoting competitiveness. Overall conclusions follow in the final section.

Some differences between Japanese and British knowledge creating systems

Industrial policy, bureaucracy, and competition

Although US demands for Japan to inaugurate trade relations – delivered by Commodore Perry in 1853 – ended over 200 years of seclusion and precipitated the 1868 Meiji Restoration that many take to be the birth of 'modern Japan,' Mason and Caiger (1972: 217) argue that the change was neither a Norman Conquest nor a French Revolution. Long-established mechanisms of government helped the Meiji administration to orchestrate a rapid and coordinated assimilation of Western technology (Wolferen 1990: 493). When the National Diet opened in 1890, Meiji oligarchs created a weak parliament and sought to counterbalance it with a strong bureaucracy, staffed by their own supporters (Johnson 1982: 37). Whereas British parliamentary democracy pre-dated industrial society, the Diet's role emerged in the 'slip-stream' of Japan's industrialization and has been more concerned with ratifying rather than initiating policy. In Johnson's words, Japan is a 'plan rational' or 'developmental' state, in which government policy has shaped patterns of innovation pursued by private companies (Johnson 1982: 19). By contrast, the impetus for competitive innovation in Anglo-American capitalism generally comes from firms seeking to exploit technology in the pursuit of competitive advantage. The foremost concern of government is mainly with the forms and procedures of competition, as opposed to determining what industries should or should not exist.

After the Second World War, Japan's ability to operate as a 'plan rational' state grew considerably when the allied reformers removed military interests and freed the civilian bureaucracy from its greatest rival (Johnson 1995: 29). Meanwhile, other features of the system remained. For example, US attempts to dismantle the 'financial cliques' or *zaibatsu* and prevent their re-emergence proved to be of limited effect. After the American Occupation ended in

1952, MITI introduced measures to dilute the US inspired Antimonopoly Law and intervened to offer companies 'administrative guidance.' By 1953, MITI was calling for the 'Keiretsufication' of industry (Miyashita and Russell 1996: 37) and paved the way for the subsequent formation of the 'big six' horizontal *keiretsu* but, in each case, structured around a bank and a powerful trading company instead of a *zaibatsu*-style family holding company.

Along with horizontal groupings, Japan's industrial structure contains vertical *keiretsu*, involving sub-contracting hierarchies (in either production and distribution) that descend from a single powerful company. Long-term trading relationships typically mean that successive layers in the subcontracting hierarchy depend on orders from the company above. Top companies might also be members of horizontal *keiretsu*, reinforcing the system's stability with a three-dimensional 'grid-like' structure. In contrast to Britain's active takeover market (Wickens 1995: 240), cross share holdings across Japan's grid structure mitigate against mergers and enhance stability by knitting the system together. Spin-off companies are also rare since lifetime employment mitigates against labor mobility.

Japan's stable economic structure features a distribution of power across the industrial, bureaucratic and political sectors, which makes it difficult for any one interest group to establish a clear advantage. Like a group of swimmers treading water, major firms are often able to improve their 'fitness' (as might be indicated by sustained increases in relative market shares) but are not usually able to force their competitors out of business nor move toward monopoly by taking over rival companies. Government policy has often tried to raise Japan's 'water level,' compared to that of other countries, by helping whole industries to rise up together through even-handed support to 'clienteles' of leading firms and providing protection from the potentially devastating effects of foreign competition.

Lifetime employment and Japanese-style organizational knowledge creation

Within Japan's macro-level stability, expectations of lifetime employment provide a stable environment for knowledge creation within Japanese organizations. While the lack of labor mobility deprives Japanese organizations of a major source of new information, companies compensate for this by making every effort to marshal internal information to support effective knowledge creation within organizational boundaries. In particular, lifetime employment provides scope for extensive transactions in tacit knowledge amongst colleagues who might spend years working together. This tacit understanding supports mutual trust and helps employees to understand their colleagues' problems, together with the type of information that would be of value to them.

Although lifetime employment does not apply to everybody, knowledge creating individuals in leading companies normally work under the assumption that there is a mutually binding loyalty between them and their employer. Organizational boundaries tend to resemble 'fortress walls' that define an arena for in-house knowledge creation. This proposition does not deny the importance of external information transactions, but aims to stress the mobility associated with in-house information. Within Japan's fortress companies, huge flows of information flow between trusted colleagues. MacDonald (1995: 557–8) notes that Japanese firms long ago discovered that information used in innovation comes from many parts of the firm and not just those parts with a specific responsibility to be creative. While these flows of information might conflict with Western ideas of 'efficiency,' they allow immense flexibility within the organization. They also act against attempts to 'breach the corporate walls' in support of short-term knowledge-creating collaborations with competitors.

Nonaka and Takeuchi (1995: 242) argue that Japanese-style organizational knowledge creation resembles a 'rugby style' exchange of tacit understanding amongst individuals that move forward together as a team, for example, by traveling with a project when it moves from R&D through to marketing. On the other hand, Western-style knowledge transfer operates more like a 'relay race,' in which passing the 'baton' – in the form of packages of explicit knowledge – substitutes for flows of tacit understanding. The relay system can give R&D departments the freedom to be creative and help to avoid the dangers of 'group think,' but there is a risk of losing the baton in transfer. Moreover, this baton-transfer process can be especially difficult in the case of inter-organizational knowledge creation. For example, many Alvey projects produced high quality 'knowledge batons,' but transfer through to exploitation proved to be problematic.

Western accounts of Japan's national innovation system often ignore the significance of lifetime employment as a platform for Japanese-style organizational creation. Other writers dismiss lifetime employment as something that is on the point of disappearing as Japan somehow becomes 'more like the West.' However, the overriding point is that accepted social values discourage employees from leaving employers to pursue better opportunities elsewhere. The 1980s saw some hesitant beginnings of a labor market for specialists, particularly engineers and software developers, but the Federation of Economic Organizations (Keidanren) made it clear that this was undesirable (Wolferen 1990: 214) and the idea of lifetime employment continues to be a pivotal feature of Japan's knowledge creating system.

Japanese companies typically make long-term investments in developing the careers of their knowledge creating employees. Since lifetime employment means that today's recruits will have to solve tomorrow's problems, labor flexibility tends to be more important than specialist skills. For example, the high degree of specialization associated with a doctorate can make it difficult for firms to integrate PhD holders into their knowledge creating system. Companies generally prefer to recruit bachelor or master level graduates and studying for a PhD can seriously damage the chances of being recruited by a prestigious company. By international standards, Japan produces few doctorates. For example, in 1988, Japan produced only 1,717 engineering PhDs, while the number of natural science doctorates was a mere 881. In the same year, the US produced 4,187 engineering and 8,157 science PhDs. Despite its much smaller population, UK figures were comparable to US totals – with 4,500 PhDs in engineering and 7,200 in science. Figures for 1992 show that the number of Japanese natural science doctorates had increased to 1,009, while the figure for engineering was 2,362. Respective US figures for 1992 were 9,372 for natural science and 5,439 for engineering.[4]

Recruiting specialists, or entering into knowledge creating ventures with outsiders (without first building appropriate levels of mutual trust), runs against the grain of Japanese business. Instead, strategic development tends to rely heavily on internal organizational knowledge creating processes.

Japanese universities

Like its parliamentary democracy, Japan's higher education system evolved in the 'slip-stream' of industrialization. Moreover, it quickly became geared to serving the needs of industry by producing more graduates in engineering and applied technology than in science. By contrast, scientific activities at some of the older European universities pre-date industrial society and engineering studies often struggled to gain recognition in the shadow of established science faculties. In Britain, Oxford University did not set up its first chair in engineering until 1908 (Lorriman and Kenjo 1996: 42) and the status of British engineering has continued to be something of a poor relation to science.

The ranking of Japanese universities does not correspond directly to research activities nor the quality of teaching. Instead, the status of top universities relates to the extent to which top companies hire their graduates and the corresponding difficulty of their entrance examinations. One article described university research in blunt terms:

> Japan's universities are not places conducive to doing profound basic research. Most have rigid, hierarchical structures. Elderly faculty heads dominate the research programs and publishing process. Intellectual dissent is discouraged. And in addition to having their ideas stifled, younger researchers often find themselves carrying out the grunt work that is done in America or Europe by technicians. Lack of money for support staff means that many people who would be better employed doing experiments are doing the washing up instead.
>
> (*The Economist*, 25 May 1996: 102)

Although Japan leads the world in terms of per capita R&D expenditure, spending by private companies accounts for about 80 percent of this total, compared to 50 percent for Britain (Barker 1996: 9). As Porter (1990: 397) observes, Japanese companies are:

> the principal engine of research and development in Japan. University research is limited, and interchange between companies and universities is modest compared to a number of other nations.

Japan's fortress companies, their isolation from the university sector and the limited scale of university research all represent fundamental differences between Japanese and British knowledge creating systems.

The changing role of MITI's computer RAs

MITI's computer RAs emerged as one of several measures to support Japan's fledgling computer industry. For example, Anchordoguy (1989: 15) mentions four types of support initiatives:

1 protectionist regulation;
2 heavy subsidies;
3 the establishment of a national company to rent domestically produced computers to Japanese users at very favorable rates (This was the Japan Electronic Computer Company (JECC) established in 1961 to buy systems from Japanese suppliers and rent them to domestic users at subsidized rates); and
4 cooperative R&D projects.

Although cooperative research programs – and the idea that synergy might offer 'something for nothing' – eventually caught the West's imagination, this conversion followed nearly two decades of MITI computer RAs.

In the 1960s and 1970s, MITI acted as a kind of 'honest broker,' using RAs to give relatively balanced subsidies to a small clientele of leading computer firms. As Samuels (1994: 69) points out:

> Six firms dominate the Japanese computer industry: Fujitsu, Hitachi, NEC, Toshiba, Mitsubishi Electric, and Oki Electric. These same firms have participated in virtually every MITI-sponsored venture since the mid-1960s and what is more they have been the *only* participants in these ventures.

MITI's long-term dialogue with its clientele of computer makers supported the gradual evolution of a series of 'modular' RAs that aimed to catch-up with America's IBM. During the catch-up era, MITI and the participating firms shared the costs of successive computer RAs. These began with the 1962–66 Computer Basic Technology Research Association involving Fujitsu, NEC and Oki.

Next came MITI's RA to develop a Very High Speed Computer System, which stretched over a six-year period from 1966 and included all six major manufacturers. While IBM continued to maintain a lead despite steady progress on the Japanese side, MITI's 1972–76 New Series Project was something of a turning point. Japanese producers divided the market and collectively provided an across-the-board challenge to IBM.

Consolidation came with MITI's celebrated 1976–80 VLSI project. This included five of the big six firms. It excluded Oki, which had failed to exploit outputs from the New Series Project and had financial problems. MITI insisted on having a single-site cooperative laboratory, but according to Okimoto, Sugano and Weinstein (1986: 561), its work concentrated on technologies that only accounted for a minor proportion of the project's overall research activities. While the firms were reluctant to compromise their independence, IBM represented a powerful common enemy and they accepted an element of common working as the price of MITI support (Callon 1995: 85–114). However, the bulk of the project concentrated on modular near-market support to accelerate the firms' development efforts. Okimoto (1986: 561) comments that:

> While the VLSI project did advance the state of Japanese semiconductor knowledge, especially in the area of production technology (e.g., silicon crystal growth and processing), Japanese companies probably would have made such advances anyway. If so, the project's main achievement may have been to hasten the timetable of development, a non-trivial but hardly revolutionary accomplishment.

The VLSI project marked the culmination of MITI's catch-up efforts and thereby helped to bring so-called cooperative research to the West's attention. However, in the words of Irving Berlin, 'the toughest thing about success is that you have to keep on being a success.' In the event, MITI's vision of future success had far-reaching repercussions.

The FGCS project involved eight firms (the six 'regular' members plus Matsushita and Sharp), together with Nippon Telegraph & Telephone (NTT). It aimed at unified joint knowledge creation centered on a common research facility: the Institute for New Generation Computer Technology (ICOT). Unified cooperation and an apparent lack of immediate commercial applications amplified participants' misgivings about working together. Cusumano (1992: 411) comments that, of the participating firms, only the company that agreed to produce the hardware showed enthusiasm. Although MITI originally planned that the firms should contribute 50 percent of project costs, the eventual 54.2 billion Yen budget (for the period 1982–92) came entirely from the government. (The Japan Information Processing Development Center, which is a semi-public corporation supported by MITI, worked to diffuse project results until 1994.)

Assessments of FGCS's achievements vary according to the weight attached to original expectations. In 1992, MITI hosted an international conference in Tokyo to report on project results, but was unable to demonstrate the achievement of the lofty goals of artificial intelligence mentioned at the outset. While the project sparked an international boom in artificial intelligence research, this peaked in the mid-1980s and had collapsed before MITI's 1992 conference. Other achievements cited include training and Cross (1992: 13) reports that, over ten years, a total of 184 young researchers had experience of working at ICOT. But problems stemmed from a divergence between the project's rigid agenda and the diverging interests of participating firms (Washington Post, 2 June 1992: C10). According to Nikkei Business (29 June 1992: 20–21), there had been growing tensions between the government and participants as technologies, not covered by the project, proved to be of greater-than-expected significance. It suggested that, from the firms' point of view, the value of national projects ended when Japan's catch-up era drew to a close.

By contrast with the unblushing claims made at the launch of the FGCS project, MITI's launch of its successor Real World Computer Program (1992–2002) was distinctly low key. The new project proposed working toward a flexible information system with an intuitive information processing capacity that is similar to that of human beings, but through basic research with no intention of even building a prototype computer (Tokyo Business 1993: 3). By 1994, the project had initiated research at MITI's Electrotechnical Laboratory and a partnership including sixteen Japanese firms and four foreign research institutions. MITI plans to provide 90 percent of the total budget, estimated at 60 billion Yen over ten years. Unlike FGCS's relatively fixed agenda, the project features a mid-term review to appraise progress and concentrate on more promising technologies. At a more general level, it reflects Japan's increasing commitment to basic research.

A new era for Japanese science policy

Amid international pressure for Japan to make a greater contribution to global progress in science, the Japanese government has been increasing budgets for basic research and promoting links with overseas research efforts. For example, at the 1987 Venice Economic Summit, Japan proposed the Human Frontier Science Program (HFSP) to conduct fundamental research in biological functions. Following international consultation, HFSP began operation in 1989 from headquarters in Strasbourg, France. To secure a timely start, Japan agreed to contribute much of the budget for an initial three-year period. In 1994, HFSP's total budget was $37 million, of which

80 percent came from Japan – with Canada, France, Germany, Italy, Switzerland and the UK providing the balance. Big science is another area where Japan is expanding its commitment to international research (Normile 1996: 1046–48). For example, when SPring-8 (Super Photon ring 8GeV) begins operation in 1997, it will provide the world's most powerful ultra-high brilliance X-ray synchrotron radiation facility and be open to users from Japan and overseas. A further dimension to internationalization is evident in Japan's moves to introduce fellowship schemes that allow foreign scientists to conduct research within Japan. These typically include the support necessary to enable overseas scientists to work effectively with Japanese researchers despite language and cultural differences. In November 1995, the Japanese government enacted a Science and Technology Basic Law that aims to ensure adequate public funding for research. Japan is also seeking to increase its production of doctorates and has introduced plans to raise the number of post-doctoral fellowships (which were rare in the early-1990s) to 10,000 during the period 1998–2000.

For Japan, it seems that relatively unified collaborations – such as MITI's Real World Computer Program – are consistent with a larger role in the international production of basic knowledge. Whereas applied R&D was of overwhelming value during the catch-up era, there is a parallel need to cover the moving frontier of international progress in more basic research. Cooperation in support of basic research could help compensate for Japan's otherwise limited 'boundary spanning' (i.e., 'inter-organizational') knowledge creation, for example, between industry and academia. However, the issue of effective 'gatekeeping,' capable of integrating significant flows of external knowledge into intra-organizational knowledge creation, requires careful attention (MacDonald and Williams 1994: 123–32).

Britain's response: The Alvey Programme

Whereas MITI's relationship with its small clientele of computer firms evolved gradually, Alvey attempted to establish a much more complex series of collaborations almost overnight. It combined funding from three different sections of government – the Department of Trade and Industry; the Ministry of Defense, and the Science and Engineering Research Council (SERC). Overall management was provided by a specially formed 'Alvey Directorate,' staffed by representatives of the sponsoring departments, as well as industry and academia. Alvey involved no less than 127 firms, together with 74 universities and polytechnics, working on a total of 192 collaborative projects (typically involving both firms and academic institutions in three-year efforts to foster unified knowledge creation), as well as 117 projects in which academics worked under the supervision of an industrial 'uncle.'

Alvey's total budget was £350 million, £200 million of which came from the three government agencies, with industry providing the remainder.

Industrial participants received 50 percent of their costs from government, while academics were funded in full by SERC. Projects covered a wide spectrum of IT technologies grouped under VLSI, software engineering, intelligent knowledge-based systems and man-machine interfaces – and covered areas where Japanese RAs had met with problems. For example, commenting on MITI's RAs in software engineering, Cusumano (1992: 389) notes a tendency toward: 'poor planning, disagreements on objectives and poor results.'

A striking feature of Alvey was that participants generally shared the government's belief in the synergic benefits of Japanese-style collaboration. Alvey research projects involved more than 5,000 researchers, who were mostly eager to 'make their projects work,' but had to learn how to assemble and maintain collaborations amidst a rapidly changing IT industry. By contrast with Japan's stable industrial structure, the British IT industry was in a state of flux and Alvey's 'batons of knowledge creation' were often lost-in-transit. Mergers and takeovers frequently disrupted collaborative groupings, while shifting commercial priorities sometimes prompted top management decisions to withdraw from projects that were technically successful. The apparent relevance of a project could all too easily diminish during the elapsed time taken to assemble a project consortium, agree on a research agenda, receive official approval and complete the work program.

An independent evaluation of Alvey concluded that pre-competitive R&D programs are well-suited to a range of tasks but are not in themselves sufficient to bolster competitive performance. Complementary private sector and government sector initiatives are needed to relate IT development to users, and to promote effort within firms to formulate strategies to facilitate the exploitation of research. Furthermore, there was a need for a serious re-evaluation of the requirement of patient capital (Guy, Georghiou, Quintus, Cameron, Hobday and Ray 1991: vii).

Even though Alvey-style collaboration did not appear to be an entirely satisfactory solution to the problem of revitalizing the UK IT industry, collaborative research formed the backbone of the successor Information Engineering Advanced Technology Program (IEATP). This retained a broadly similar structure to Alvey but funding was much more modest. Half of its approximately £100 million budget came from the DTI and SERC, with industry providing the balance. IEATP also differed from its predecessor in giving preference to small and medium sized enterprises. It placed more emphasis on routes to exploitation and attempted to increase the level of user involvement in the program. In commenting on the evaluation of Britain's IEATP Programme, Buisseret and Cameron (1994: 222) note that project managers consistently favored collaborations that involved a user organization, that posed a 'problem,' an organization that provided a 'solution' and a research organization (usually an academic institution) that contributed additional expertise. Such a model appears to involve modular rather than unified knowledge creation.

A new direction for UK IT support policies followed from the UK government's White Paper (1993), which stressed the need for 'wealth creation' through fostering the climate for innovation, developing access to technical help and facilitating technology transfer. To support these grand goals, the White Paper advocated the use of a 'Technology Foresight' (TF) exercise to identify 'generic' technologies that are likely to yield the greatest economic and social benefits to the UK in the long term. The argument is that Foresight can inform decisions by government, industry, research councils and academia on R&D priorities and underlying skill needs. Foresight followed an examination of overseas experiences with Japan being cited as one of three counties for special attention. In March 1994, a high-level delegation visited Tokyo, professing an admiration for Japanese TF and a desire to learn the lessons of Japan's success. With regard to policy issues for the UK, a subsequent report by the mission's leader Chisholm (1994: 14) reflected that:

> Japan is so different culturally and structurally from the UK that a direct read across of their experience is probably fruitless if not dangerous. However, the visit did provoke some thoughts that might be worth exploring: TF's principal pay off appears to be the effect it has in bringing coherence to the national vision of the technology future and, as such has a clear relevance to the UK at this time; Japanese companies appear to gain strength from their willingness to work together on pre-competitive research and share information.

Chisholm's initial caution about 'direct read across' seems to give way to a view that the Japanese experience – for example, in TF and pre-competitive cooperative research – might nevertheless be of clear relevance to the UK.

Why collaborate?

Collaboration can be a good servant but a difficult master. At one level, there are 'natural' research collaborations that can operate effectively to the mutual benefit of participants: for example, between a buyer and seller, cooperation to combine complementary skills or working together to establish standards. Success typically depends on everyone having something to contribute and a good reason to enter into information transactions with other partners. However, creating these qualities in 'artificial' government orchestrated collaborations can be something of a challenge. This appears to be especially true in horizontal collaborations between competitors seeking to suspend competition (to obtain economies-of-scale during pre-competitive cooperative research) before attempting to exploit results competitively. MacDonald (1992: 50–51) sums up the problem of pre-competitive collaboration succinctly:

> To help identify the sort of collaborative activity that they are anxious to encourage, governments – and firms anxious to be encouraged – have construed the notion of 'pre-competitive' research as investigative activity undertaken by firms to provide knowledge which is not specifically relevant to their competitive positions. The concept is obviously akin to the more traditional one of 'basic' research – providing the knowledge infrastructure on which applied research can build – but it suffers from a more severe form of the same deficiency. While it is just possible to imagine firms being willing to contribute to a common pool of knowledge from which they also wish to draw, it strains credibility to think that such activity has nothing whatsoever to do with competition. The firm's ability to compete determines whether it will survive and prosper; there is no part of its operations – certainly not R&D – which is totally removed from the need to compete.

As Macdonald points out, pre-competitive research resembles the idea of support for basic research. Paradoxically, both the FGCS and Alvey projects demonstrated that the greatest achievements of such programs are likely to be closer to the idea of reinforcing basic research infrastructures than providing direct enhancements to competitiveness. Nevertheless, Japan and Britain reacted to their respective experiences by opting for two very different types of follow-on program.

Conclusion

Throughout its catch-up era, Japan proved to be effective at abstracting Western technology and transplanting it within the Japanese system. Nevertheless, the country's national innovation system continued to operate in a distinctly Japanese way. Unlike Britain's Industrial Revolution, in which turbulent entrepreneurial capitalism led the world into an industrial age, the Japanese government orchestrated the birth and development of evolutionary industrialization. The long-standing legitimacy of government intervention in Japanese industry (together with associated channels of communication) is markedly different from British concerns about the nature and extent of acceptable intervention in the 'market rational' aspects of its innovation system.

Japan's government-coordinated industrialization supported rapid advances toward 'modern nation' status and an ability to command Western respect. However, simply developing the outward signs of Western democracy does not necessarily mean that the underlying system becomes correspondingly less Japanese. For example, it is misleading to assume that Japan's parliamentary democracy and university system replicate the operation of their British counterparts. Overseas policy-makers seeking to abstract lessons from

Japan often suffer from the problems of trying to peer through a two-way mirror from the wrong side. While they might see blurred images that apparently equate to familiar features of their own national innovation systems, they often fail to grasp the underlying processes of Japan's 'plan rational' system.

Japan's post Second World War recovery emphasized the country's 'plan rational' capability to make spectacular progress in the face of an overwhelming need to rebuild the economy and catch up with best-practice Western technology. At a time when Britain tended to place faith in economic growth through science-related areas, Japan's private companies sought to replicate Western achievements in applied technology. They did this within a remarkably stable framework that combined competition with government measures to help the whole system catch-up (i.e., raising the 'water level' relative to other countries). More recently, international pressure to make a greater contribution to progress in science has been instrumental in encouraging Japan to complement its established strengths in applied R&D with larger budgets for public sector research.

MITI's succession of catch-up computer RAs, in the 1960s and 1970s, supported near-market development by channeling subsidies to leading computer makers. While the policy offered a mechanism for balancing subsidies according to changing circumstances (the role of an 'honest broker'), knowledge creation within the RAs does not appear to have been especially 'collaborative' (i.e., it was more modular than unified). Moreover, IBM provided a 'guiding light' that illuminated paths of development toward technologies of proven commercial value. MITI's computer RAs (together with a raft of supporting policies) were effective in helping to close the gap but their effectiveness raised the issue of what to do next. In the event, FGCS's high-profile launch, together with Western belief in both its potency and the value of pre-competitive collaborative research, had far-reaching consequences.

Neither the FGCS project nor Alvey provided strong evidence to support the use of 'pre-competitive' collaborative research as a direct route to enhanced competitiveness. The value of such projects appears more akin to the traditional role of basic research and activities, such as standards development, where there is a 'natural logic' for a cooperative approach. Having suffered from the problems of giving 'hostages to fortune' with the FGCS program, MITI appeared to accept the natural limitations of 'pre-competitive' cooperative research in its claims for the Real World Computer Program, which it presented as avowedly basic research involving partners from Japan and across the world. Given the limited nature of Japanese university research and the reluctance of 'fortress companies' to become involved in projects that could compromise their independence, basic research RAs provide one route to encouraging inter-organizational knowledge exchanges.

By contrast with Japan, Britain has long-established strengths in basic research. Moreover, in 1988 (the last year of Alvey) its production of natural science doctorates was more than eight times Japan's total, despite having less than half the population. While Alvey made great progress in unifying the disparate aspects of Britain's IT knowledge base, it seems extravagant to expect that this type of collaboration can easily contribute directly to enhanced competitiveness. Retaining collaboration for the Information Engineering Advanced Technology Program (see above), albeit with more emphasis on exploitation, implies a peculiar degree of residual faith in the FGCS-inspired 'collaborate-to-compete' model.

Alvey differed from MITI's computer RAs by aiming to implement a novel and enormously complex venture using a specially created organization in the midst of tremendous commercial turbulence. The difficulties of assembling projects, overcoming the often cumbersome challenge of integrating academic components into industrial-style work-schedules and developing appropriate project management procedures were substantial. Even then, factors – such as rapid technological developments, shifting business strategies, structural changes in the IT industry and concerns over the ownership of intellectual property – could all frustrate the exploitation of an apparently well planned project. Pre-competitive collaborative research seems to be inappropriate as a policy tool for directly promoting increased competitiveness. Although simplifying the collaborative model to the point where it resembles a 'natural collaboration' might increase the probability of exploitation, the result can be closer to a near-market subsidy for modular work than unified inter-organizational knowledge creation.

In a manner not altogether dissimilar from Alvey's background, Britain's TF exercise sought to learn from the Japanese experience. However, irrespective of the merits of TF, it seems that there are good reasons for being cautious in using Japan as a model for UK policy. Since Japanese and British national innovation systems have developed differently and operate differently, there is a clear need to take due account of the nature and implications of these differences in attempts to learn from Japan.

NOTES

1 The argument draws partly on the author's experience as a member of a team – at the Programme of Policy Research in Engineering, Science and Technology (PREST), University of Manchester, UK – that evaluated the structure and organization of the Alvey Programme and its successor, the Information Engineering Advanced Technology Programme (IEATP). However, views expressed in the chapter do not represent official evaluation findings, which are published separately (Guy, Georghiou, Quintus, Cameron, Hobday and Ray 1991; Cameron, Georghiou, Buisseret and Ray 1995). The chapter also uses results of thirty-four interviews conducted in Japan with policy-makers, senior industrialists and researchers involved with MITI's collaborative research schemes, undertaken

during a Science and Technology Agency Fellowship (1992–93) at the National Institute of Science and Technology Policy (NISTEP), Tokyo.

2 MITI's RAs were loosely based on a British system (that later spread to other European countries) introduced after the First World War amid fears that the UK was losing its competitiveness. British Research Associations, financed by the government and private firms, were established to provide overall support for an industry. Japan passed enabling legislation for RAs in the 1961 Act on Mining and Manufacturing Industry Technology Research Association. Odagiri and Goto (1993: 87) note that Japan organized a total of eighty-seven RAs during the period 1961–87.

3 Brian Oakley (1994: 2), who was the Alvey Programme's original Director, offered this speculation in comments about a decade of European cooperative research schemes. While acknowledging the many positive effects of the programs he noted that, in the limit, it is difficult to justify them if their original objective – to enhance the prosperity of Europe's IT industries – was not fulfilled.

4 PhD data sources: Unpublished tables provided by Japan's Ministry of Education, Science, and Culture and the US National Science Foundation. Note that in the Japanese data, mathematics is included in natural sciences, while computer science is grouped under engineering.

REFERENCES

Anchordoguy, M. (1989) *Computers Inc.: Japan's Challenge to IBM*. Cambridge, Mass.: Harvard University Press.

Barker, B. (1996) *Japan: A Science Profile*. UK: The British Council.

Buisseret, T. and H. Cameron (1994) Management of Collaborative Research: Collaboration and Exploitation under the UK's Information Engineering Advanced Technology Programme. In *Technology Analysis & Strategic Management*, vol. 6, no. 7, pp. 215–30.

Callon, S. (1995) *Divided Sun: MITI and the Breakdown of Japanese High Technology Policy*. Stanford: Stanford University Press.

Cameron, H., L. Georghiou, T. Buisseret, and T. Ray (1995) *The Evaluation of the Information Engineering Advanced Technology Programme: Final Report*, London: HMSO.

Chisholm, J. (1994) *Turning Research into Wealth: The Japanese Way*, A Foresight article. London: HMSO.

Cross, M. (1992) Farewell to Computing's Fifth Generation. In *New Scientist*, vol. 13, June, pp. 12–3.

Cusumano, M. (1992) *Japan's Software Factories: A Challenge to US Management*. New York and Oxford: Oxford University Press.

The Economist (1996) *Science and Technology: Back to Basics in Japan*. 25 May, p. 102.

Evan, W. and P. Olk (1990) R&D Consortia: A New US Organizational Form. In *Sloan Management Review*, vol. 1, no. 3, pp. 37–46.

Ferné, G. (1989) R&D Programmes for Information Technology. In *The OECD Observer*, August/September, pp. 10–3.

Guy, K., L. Georghiou, P. Quintus, H. Cameron, M. Hobday, and T. Ray (1991) *Evaluation of the Alvey Programme for Advanced Information Technology*. London: HMSO.

Johnson, C. (1982) *MITI and the Japanese Miracle: The Growth of Industrial Policy, 1925–1975*. Tokyo: Charles E. Tuttle.

—— (1995) *Japan: Who Governs? The Rise of the Developmental State*. New York and London: W. W. Norton and Company.

Lorriman, J. and T. Kenjo (1996) *Japan's Winning Margins*. Oxford: Oxford University Press.

MacDonald, S. (1992) Formal Collaboration and Informal Information Flow. In *International Journal of Technology Management, Special issue on Strengthening Corporate and National Competitiveness through Technology*, vol. 7, no. 1/2/3, pp. 49–60.

—— (1995) Learning to Change: An Information Perspective on Learning in the Organization. In *Organization Science*, no. 62.

MacDonald, S. and C. Williams (1994) The Survival of the Gatekeeper. In *Research Policy*, vol. 23, pp. 123–32.

Mason, R. and J. Caiger (1972), *A History of Japan*. Tokyo: Charles E. Tuttle Company.

Miyashita, K. and D. Russell (1996) *Keiretsu: Inside the Hidden Japanese Conglomerates*. New York: McGraw-Hill.

Nikkei Business (1992) *Tokushū: fukuro-kōji no gijutsu taikoku – kōryoku ushinau kanmin kyōdō kaihatsu – 'oitsuke oikose' jidai no ibutsu dai go sedai konpyūtā shōgyōka tōku* (Special feature: Japan's technological dead-end – Government-public cooperative development [projects] are losing their effectiveness – The Fifth Generation Computer is a long way from commercialization, a remnant of [Japan's] 'catch-up, overtake' era), 29 June, pp. 20–1.

Nonaka, I. and H. Takeuchi (1995) *The Knowledge Creating Company*. New York and Oxford: Oxford University Press.

Normile, D. (1996) Big Science is Booming in Japan. In *Science*, vol. 271, 23 February, pp. 1046–8.

Oakley, B. and K. Owen (1989) *Alvey: Britain's Strategic Computing Initiative*. Cambridge, Mass.: MIT Press.

Oakley, B. (1994), Idle Thoughts of an Old Hand. Paper presented at a conference on '*Management of Collaborative European Programmes and Projects in Research, Education and Training*'. Oxford, 10–13 April 1994.

Odagiri, H. and A. Goto (1993) The Japanese System of Innovation: Past, Present and Future. In Nelson, R. (ed.), *National Innovation Systems*. New York and Oxford: Oxford University Press, pp. 76–114.

Okimoto, D. (1986) The Japanese Challenge in High Technology. In Landau, R. and N. Rosenberg (eds), *The Positive Sum Strategy: Harnessing Technology for Economic Growth*. Washington, DC: National Academy Press, pp. 541–67.

Okimoto, D., T. Sugano and F. B. Weinstein (1986) *Competitive Edge: the Semiconductor industry in the USA and Japan*. Stanford, Ca.: Stanford University Press.

Porter, M. (1990) *The Competitive Advantage of Nations*. New York: The Free Press.

Samuels, R. (1994) *Rich Nation Strong Army: National Security and the Transformation of Japan*. New York: Cornell University Press.

Tokyo Business (1993) *Real World Computing: A Brave New World*, March, p.3.

Washington Post (1992) *Japan's 'Fifth Generation' Computer Development Project Comes to an End*, 2 June, p. C10.

White Paper (1993) *Realising our Potential: A Strategy for Science Engineering and Technology*, London: HMSO, Cm 2250.

Wickens, P. (1995) *The Ascendant Organization*. London: Macmillan.
Wolferen, K. V. (1990) *The Enigma of Japanese Power: People and Politics in a Stateless Nation*. London: Macmillan/PAPERMAC.
Woodward, F. (1965) *Structure of Industrial Research Associations*. Paris: OECD, Directorate for Scientific Affairs.

Part IV

MAJOR FIELDS OF SCIENCE AND TECHNOLOGY

10

R&D IN JAPAN'S PHARMACEUTICAL INDUSTRY

The biological revolution, gene therapy and public policy

Christian Oberländer

Introduction

Opening of the Japanese pharmaceutical market and cost-containment in the national health care system are confronting Japan's pharmaceutical industry with an existential challenge. For the preservation of this industry, the creation of independent research capabilities in future key technologies for the development of internationally marketable products appears indispensable. The required R&D efforts coincide with the 'biological revolution,' posing additional challenges for the industry, but also opening up new opportunities.

In this paper, the general development of R&D in Japan's pharmaceutical industry and the role of government, especially of the Japanese Ministry of Health and Welfare (MHW) in supporting these R&D efforts are discussed. Then, focusing on the example of gene therapy, the instruments of Japan's research promotion for pharmaceutical industry will be examined in the light of the implications of the 'biological revolution' for the innovation strategy of this industry.

Japan's pharmaceutical R&D and MHW's research promotion policy

Challenges to the pharmaceutical industry and increasing importance of R&D

The changes in strategy of Japan's pharmaceutical companies and their R&D efforts are caused by several factors challenging the status quo of the Japanese pharmaceutical market. First, this market is increasingly opened up. When the Japanese drug market was accessible only through cooperation with domestic manufacturers, Japanese pharmaceutical companies profited from international product development through licensed production and sales tie-ups with foreign manufacturers (Yoshikawa 1989: 83). International harmonization of drug approval standards gives foreign companies now increasingly easier access to Japan's market. In April 1997, for example, a system of clinical trials following international standards was introduced. However, the growing presence of foreign companies in Japan and the subsequently increasing direct sales of their products lowers income from this source. Fujisawa's sales, for example, dropped by 17 billion Yen in 1995 after a sales agreement with SmithKline Beecham expired (*Nikkei Weekly*, 3 June 1996). The market share of foreign pharmaceutical companies has reached 25 percent in the mid-1990s and is expected to rise to 40–50 percent by the year 2010 (*Yakuji handobukku* '97: 59–60).

Second, several domestic developments have put pressure on the earnings of Japanese drug manufacturers. Particularly, MHW is intensively working toward cost control in health care. An important target of its campaigns are pharmaceutical expenditures which in Japan occupy the highest percentage in health care spending among industrialized countries. In 1993, the cost of drugs accounted for 29.6 percent of Japan's total medical expenditures, while in other industrialized countries, this percentage was significantly lower, for example 17.1 percent in Germany and 11.3 percent in the US (*Shūkan Shakai Hoshō* 1996: 26). To promote cost-containment, the ministry has developed several policy initiatives. Originally on a bi-annual basis, however since 1996 on an annual basis, it forcibly reduces the reimbursement for drugs by health insurance, thereby lowering the profit margin of manufacturers. In spring 1998 for example, according to MHW's plans, drug prices were cut by an average of about 10 percent (*Nihon Keizai Shinbun*, 26 October 1997). Furthermore, the ministry tries to separate drug prescription and dispensing which is largely unified in the hands of doctors. Since they profit directly from drug sales, the lack of separation of these two functions is seen as a main cause of over-prescription of drugs in Japan. Finally, the beginning of replacement of the present fee-for-service reimbursement through flat-fees is likely to suppress the consumption of drugs even further (Kimura *et al.* 1993: 184). These pressures on Japan's pharmaceutical industry are expected to intensify in the future. MHW is preparing for health care reform that would introduce a new system of reimbursement for drugs which would largely eliminate incentives for generous prescriptions. The application of flat-fees would also be expanded greatly (*Nihon Keizai Shinbun*, 22 September 1997).

A possible way of coping with declining sales in Japan's domestic market would be the development of products which could be sold internationally. This would allow Japanese drug manufacturers, whose degree of internationalization is still very low, to make up for lost earnings in the domestic market through international sales. This is only possible by undertaking great and risky R&D efforts. However, at a time when the development of new drugs is viewed by many no longer as a lottery, but as influenced by the size of the investment made, the rapidly rising cost of R&D in this sector combined with the smaller sales base of Japanese pharmaceutical companies put limits on their R&D capabilities. From a sales base and in an international perspective, Japanese companies are only of medium size: Takeda as the largest Japanese company ranks 13, followed by the companies Sankyo and Yamanouchi which rank 18 and 20 respectively (Table 10.1). Therefore, even if Japanese manufacturers greatly increase their percentage of sales spent on R&D, their smaller sales base hinders them in keeping up with the absolute research expenditures of the international pharmaceutical giants (Sapienza 1993).

MHW's research promotion policy for the pharmaceutical industry

Reich points out that until the 1980s, 'the Japanese government did not develop a clearly articulated sector-specific industrial policy for the pharmaceutical industry' (1990: 145). However during the early 1980s, when the need to support the pharmaceutical industry was felt keenly, MHW took on the difficult task to pursue cost reduction in health care and thereby weakening

Table 10.1 Comparison of the world-wide largest pharmaceutical companies (1995, million $)

Rank	*Name*	*Sales*	*World Market Share (%)*	*R&D Exp.*	*R&D /Sales (%)*
1	Glaxo Wellcome	12,586	4.4	1,894	15.0
2	Merck (US)	11,314	4.0	1,331	11.8
3	Novartis	10,571	3.7	1,691	16.0
4	Hoechst-Marion-Roussel (HMR)	8,438	3.0	1,200	14.2
5	Bristol Myers Squibb (BMS)	7,810	2.7	1,007	12.9
6	Pfizer	7,072	2.5	1,295	18.3
7	AHP	7,005	2.5	1,100	15.7
8	Roche	6,800	2.4	1,660	24.4
13	Takeda Chemical	5,621	2.0	657*	8.7*
18	Sankyō	4,353	1.5	463*	8.6*
20	Yamanouchi	3,576	1.3	265*	9.9*

Sources: Scrip Magazine, January 1997, cited according to *Yakuji handobukku* '97 (1997: 111–12); figures marked with asterisks cited according to *Yakuji handobukko* '96 (1996: 118–19).

the domestic position of Japan's pharmaceutical industry while at the same time trying to promote the industry's international competitiveness. The Pharmaceutical Industry Policy Forum (*Iyakuhin sangyō seisaku kondankai*) that was set up by MHW in 1982 confirmed the importance of an increased R&D effort for the future of Japan's pharmaceutical industry. Research and development of innovative products, for example in biotechnology, appeared to be a key to the industry's future prosperity. The Forum formulated 'Strengthening R&D and expanding in overseas markets' as the guideline for policies to support the future development of Japan's pharmaceutical industry (quoted according to Reich 1990: 138).

In 1983, an Office for the Promotion of Advanced Technology for Pharmaceuticals (*iyakuhin sentan gijutsu shinkō shitsu*) was established within the ministry's Bureau for Pharmaceutical Affairs (*yakumukyoku*) which was in charge of policy for the pharmaceutical industry. In July 1994, this Office was promoted to become the Division for Research Promotion (*kenkyū kaihatsu shinkōka*) of the Department for Pharmaceutical Affairs. It was staffed with about fifteen people and responsible for the support of the research efforts of Japan's pharmaceutical industry. This Division also coordinated the activities of the two most important external bodies (*gaibu dantai*) for research promotion, the Health Sciences Foundation and the Drug Organization.

Following the recommendations issued by the Pharmaceutical Industry Policy Forum in 1984, the predecessor of the Japan Health Sciences Foundation (HSF) was founded in 1985, only to take on its present name in April 1986. This foundation which is legally independent but which is supported by the ministry and about 150 member companies, has the function to promote research cooperation between universities, national research institutes and industry in the area of medical and pharmaceutical basic research (*Human Science* 1995: 31, 67, 69). In 1992, the foundation had a budget of about 1.5 billion Yen for research promotion. Besides disseminating information and conducting seminars on current research topics, for example 'From Genetic Information to Drug Discovery' (November 1996; *Human Science* 1997: 32–3), this organization sponsors different types of research. At the core of the foundation's activities are the Integrated Study Projects in Drug Innovation Science which aim at the creation of basic technology for innovation in pharmaceuticals. Besides this, the foundation has specialized programs which also fund applied research. For example, the search for drugs to combat AIDS is supported with a special budget of 400 million Yen. In addition, the foundation functions as an advisory body to the ministry in the drafting of norms and regulations by utilizing its close contacts to industry (Koseisho Yakumukyoku 1993: 54–5).

Also based on the recommendations of the Pharmaceutical Industry Policy Forum, changes were made in the structure of the Organization for Adverse Drug Reaction Relief, R&D Promotion and Product Review ('Drug Organization') whose activities were expanded to support the entire process from basic research to development and application. This organization is a special corporation (*tokushū hōjin*) which allows MHW to delegate its own staff there. It was founded in 1979 with the original aim to provide compensation to citizens who have become victims of drug scandals. However, its functions were later expanded to include research promotion which reached a value of 2.5 billion Yen in 1992. This budget is divided into capital donations and research support. Under the capital donations program, the organization can provide up to 70 percent of the required capital over a period of ten years to research consortia of private companies. Research support measures comprise the incubation of research cooperation between industry and national institutes, the implementation of projects which industry can not realize by itself, and the preparation of studies concerning the latest research and health trends (Koseisho Yakumukyoku 1993: 56–7). Both the HSF and the Drug Organization's programs put a certain emphasis on company collaboration and cooperation between business, academia and national laboratories.

Further initiatives to support research by the pharmaceutical industry were prompted by a memorandum (*kankoku*) released by the Science Council of Japan in October 1990. This memorandum On the Promotion of Basic Sciences for Drug Creation pointed to Japan's underdeveloped research infrastructure in the life sciences and led to the formation of a Study Group (*kentōkai*) for the Promotion of Drug Creation Research by the Bureau for Pharmaceutical Affairs of MHW. Following its recommendations of 1991, MHW reorganized several of its grant programs under the plan Pharma Dream 21. Additional funds were appropriated for General Research into Drug Creation Sciences and for Research into the International Harmonization of Drug Regulations (Koseisho Kosei Kagakka 1993: 65–8).

The string of bodies concerned with the pharmaceutical industry continued when, in October 1992, the Forum for the Pharmaceuticals of the twenty-first Century was established. The fact that the members of this Forum had to deal with the two issues of Research Promotion for Pharmaceuticals and the Promotion of the Correct Usage of Pharmaceuticals – a euphemism for cost-control by reducing the prescription of drugs – demonstrates again the dilemma which MHW's policy-makers are facing. The interim report of the Forum which deals with the first theme in particular, reiterated the concern about Japan's lag in basic research and the low quality of its clinical trials. However, its recommendations offered few new ideas and it appears to have had little effect in terms of MHW's funding programs (Koseisho Yakumukyoku 1993: I, 3–12).

Besides the specific measures mentioned above, MHW attempts to promote innovation in pharmaceutical research through other, indirect, mechanisms too. In the system of pharmaceutical price reduction, for example, the price of new and innovative drugs is set higher and subsequently reduced to a lesser extent than that of older drugs. However, this system is widely suspected of providing incentives for Japan's pharmaceutical manufacturers to concentrate on only slight improvements of already existing drugs in order to come out with a 'new' and more profitable product as frequently as possible (see for example Anegawa 1997: 20–3, Endo and Tanaka 1997: 71, Nambu 1997). In addition, empirical evidence even hints that the 'accumulation of price cuts [by itself] may have led to some reduction in R&D expenditure' (Mitchell *et al.* 1996: 137).

Other measures for research promotion in the pharmaceutical industry include taxation, adjustments in the drug approval process, extension of patent periods, etc. However, it has been found that these instruments are not well suited for R&D promotion (Anegawa 1997: 35). In spite of this, besides some minor changes, Japan's policy tools in this area have largely remained unchanged since the middle of the 1980s.

Has R&D of Japan's pharmaceutical industry been transformed?

The literature on Japan's pharmaceutical industry is generally critical of the degree to which MHW's policy tools have influenced the research of the industry they were meant for. Howells and Neary have pointed out that 'the amount of money made available under the new MHW policy [for research promotion] was tiny' [compared to what the industry itself is spending] and – concerning research collaborations among companies – 'even the Japanese government is unlikely to be able to encourage much meaningful research cooperation' (Howells and Neary 1995: 98, 105). In this section, the development of R&D efforts by Japan's pharmaceutical industry since the middle of the 1980s will be briefly discussed based on the data collected by the Somucho Tokeikyoku (Statistics Bureau of the Management and Coordination Agency). In particular, it will be discussed if a shift toward more R&D, especially more innovative basic R&D which MHW aimed for, actually occurred.

Figure 10.1 gives a general overview of the development of sales, profitability and R&D expenditure in the face of the reductions in drug reimbursement implemented by MHW. Following the price adjustments, sales growth has declined sharply. Profitability as the ratio of net profits to sales fluctuated more moderately, however, even less so in the first half of the 1990s. This might be because companies have learned to better cope with the periodical price reductions or because of a modification of MHW's policy which, in 1992, limited price reductions to products on which companies gave discounts exceeding 15 percent of list prices.

During the period from 1975 to 1985, Reich observes a nearly four-fold increase of corporate investment in pharmaceutical R&D and a shift in R&D strategy toward developing 'products primarily for foreign markets' (1990: 138–9, 141, 146). Since then, the growth in companies' R&D budgets has continued. Between 1984 and 1992, the R&D intensity (measured by R&D

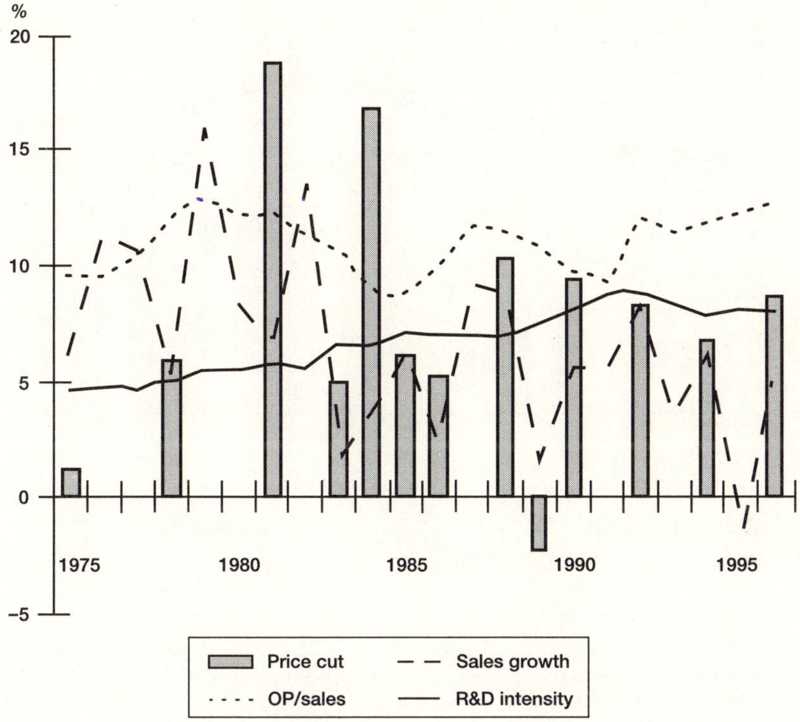

Figure 10.1 Net operating profit (OP) as percentage of sales, sales growth, R&D intensity and price cuts in the Japanese pharmaceutical industry (percent)

Source: Somucho Tokeikyoku (1996 and various years).

spending as a percentage of sales) of Japan's pharmaceutical industry grew constantly from 6.5 percent in 1984, peaked at 8.7 percent in 1992 and then stabilized around 7.9 percent by 1996 (Figure 10.1). The Japanese industry therefore still has a long way to go to reach the double-digit R&D intensity of many of the international pharmaceutical giants (Table 10.1). The findings of Mitchell *et al.* suggest that the cumulative price reductions might have affected R&D growth negatively (1996: 137).

That Japan's pharmaceutical companies have indeed expanded their R&D activities is apparent from the number of researchers employed in the industry. Their number has grown significantly, from 12,713 in 1985 to 20,041 in 1994, then declined slightly to 19,970 in 1996 (Figure 10.2). Spending of research funds per individual researcher rose from 1984 until 1992, but dipped down in 1994. It picked up again in the following years but has not yet fully recovered its former peak level. This might be an indication that the continued increase in the number of researchers in the early 1990s was caused by companies' restructuring efforts after the 'bubble economy.' At

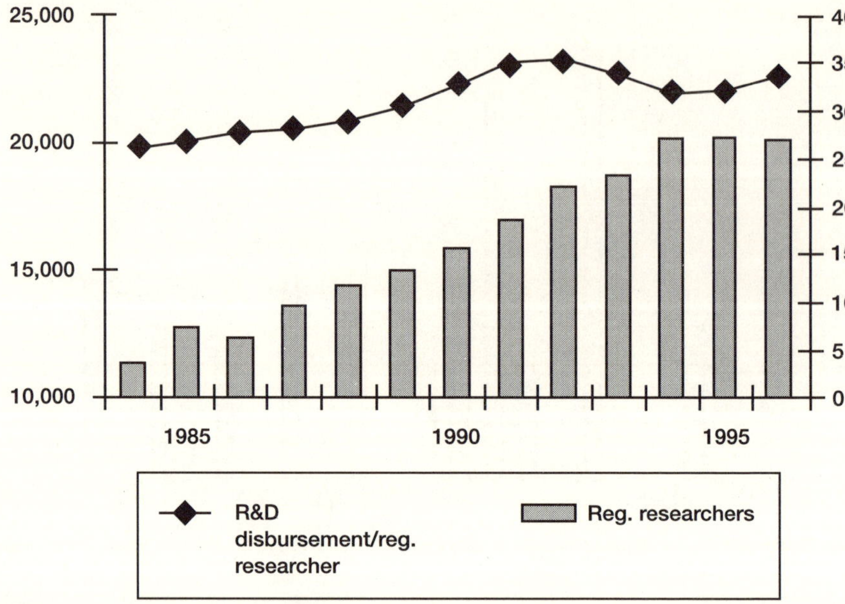

Figure 10.2 Number of regular researchers and research spending per researcher in the Japanese pharmaceutical industry (million Yen)
Source: Somucho Tokeikyoku (1996 and various years).

this time, many companies coped with excess staff by transferring personnel into research departments. That this increase in personnel was not accompanied by complementary spending might support this hypothesis (see also Chapter 8).

Although the absolute funds available for R&D and the number of researchers have almost doubled, the spending pattern of these funds seems to have barely changed. Figure 10.3 indicates that the distribution of research funds between the three areas of basic research, applied research and development was not altered substantially. In other words, the shift toward basic research that is essential for the development of innovative 'breakthrough' products has not yet occurred.

This is also reflected in the balance of trade in pharmaceutical technology. While the volume of this trade almost tripled during the ten years before 1994, imports are exceeding exports since 1991 (Figure 10.4). Japan did not manage to become a major exporter of pharmaceutical technology. Quite to the contrary, funds spend on technology imports as a percentage of R&D spending have increased during the period from 1984 to 1994 (Figure 10.5). This indicates that Japan's self-reliance in pharmaceutical technology has not increased, but has even been reduced.

Noteworthy is that technology exports to the US have grown four-fold,

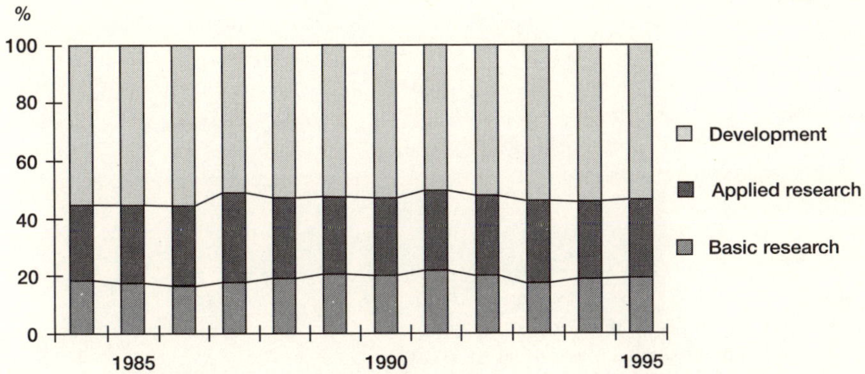

Figure 10.3 Distribution of R&D funds between basic research, applied research and development in the Japanese pharmaceutical industry
Source: Somucho Tokeikyoku (1997 and various years).

Figure 10.4 Japanese exports and imports of pharmaceutical technology (million Yen)
Source: Somucho Tokeikyoku (1996 and various years).

while those to Europe have only risen slightly. Imports of technology from the US as well as from Europe have increased; however, those from the US have shown stronger growth exceeding those from Europe in absolute value since 1991 (Figure 10.6). These figures suggest the growing role of the US as the world's 'motor' in pharmaceutical technology; a position which is probably supported by the strong growth of biotechnology in the US.

In summary, Japan's pharmaceutical industry is under pressure in its domestic market. A possible solution to this would be to increase R&D efforts, develop products that can be sold on international markets, and

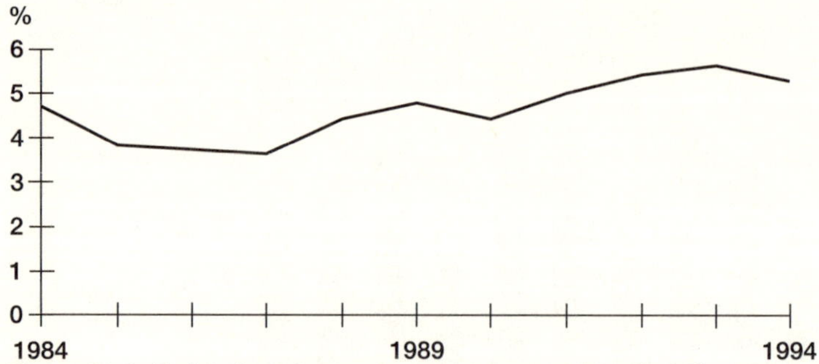

Figure 10.5 Technology imports as percentage of R&D disbursement of the Japanese pharmaceutical industry

Source: Somucho Tokeikyoku (1996 and various years).

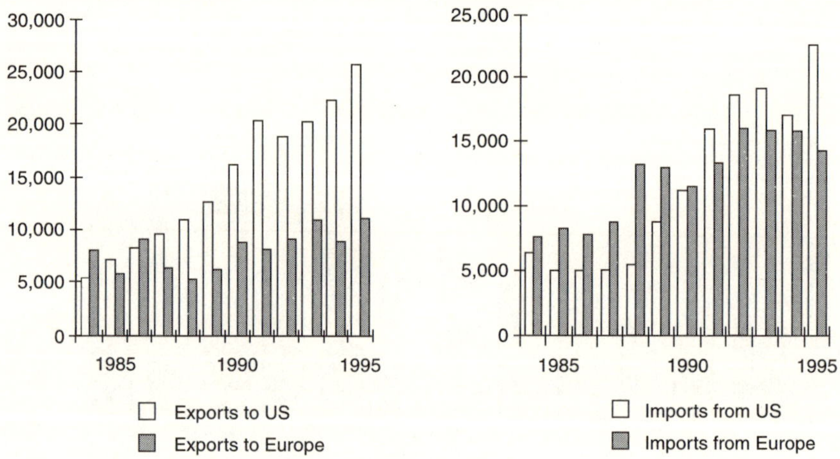

Figure 10.6 Japanese exports and imports of pharmaceutical technology to Europe and the US (million Yen)

Source: Somucho Tokeikyoku (1996 and various years).

thereby defend domestic and international market share. Japan's policy-makers at MHW have recognized this; however, their initiatives for research promotion in the pharmaceutical industry appear to have been of little effect so far. In spite of numerous support programs by MHW, increased nominal R&D spending, and an increased number of researchers, the shift to more basic research – as a general trend – has not taken place. Moreover, the fact that Japan is still importing substantial amounts of pharmaceutical technology and that its major pharmaceutical companies have not succeeded in a sustainable increase in their export shares (Figure 10.7) hints that the hoped-for innovative work has not yet taken place or has not yet taken effect.

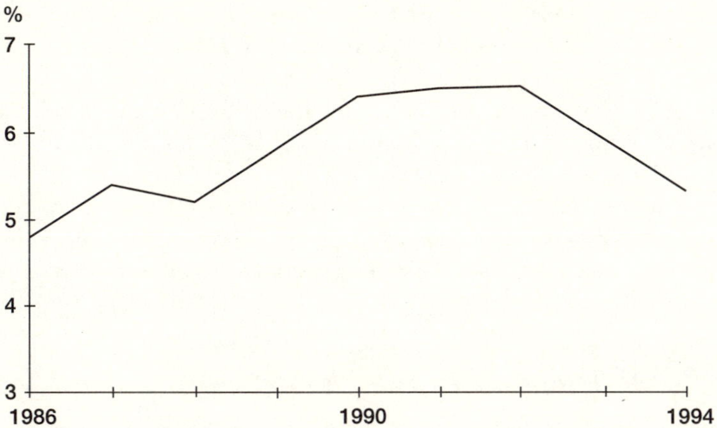

Figure 10.7 Foreign sales as percentage of total sales of Japan's twenty leading pharmaceutical manufacturers
Source: Somucho Tokeikyoku (1996 and various years).

Gene therapy – future or end for Japan's pharmaceutical industry?

The 'biological revolution' and the importance of genetics

Japan's pharmaceutical industry is not only required to expand its R&D effort but also to adjust to the new conditions created by the 'biological revolution.' Progress in biotechnology has profoundly changed the way research is conducted in the pharmaceutical industry (Della Valle and Gambardella 1993). In particular, genetic research has far-reaching consequences because understanding of the genetic foundations of a disease eventually allows the targeted development of a therapy for that illness. '"Big pharma" has come to the realization that the route to the drugs of tomorrow increasingly begins with the gene discoveries of today . . . take SmithKline Beecham (SB) . . . which . . . currently bases 25 percent of its drug discovery process on genomics . . . Moreover, . . . all of SB's research will stem from genomics by the year 2000' (*Nature Genetics* 1996: 1).

Many hope that gene therapy in particular will bring the 'fourth revolution in the history of medicine' (Zaidan hojin 1996: 1). In gene therapy, treatment is attempted by bringing genetic material with the help of a 'vector' into the nucleus of a cell to correct for example a genetic defect. The hopes for therapeutic successes outweigh fears of potential drawbacks by far and, in 1990, the first experimental gene therapy was implemented in the United States (Anderson 1995). Although final success of the trial could not yet be confirmed, this and the following more than 2,000 trials (by 1997) have established the relative safety of the procedure and opened the way for continuing clinical studies.

The potential market for gene therapy is already highly estimated. According to a forecast by Technology Catalysts, this market's future size will vary between $7 billion and $21 billion in the year 2000 and between $39 billion and $157 billion in 2010 (Nikkei Baioteku 1997: 275). This technology may therefore acquire a strategic significance for the future of the pharmaceutical industry. The US National Institutes of Health (NIH) spend every year about $200 million on the development and testing of techniques for gene therapy (Marshall 1995: 1050). The expenditures for gene therapy research made by private US companies are estimated to exceed this amount.

Gene therapy and its promotion in Japan

In Japan, gene therapy was first implemented only in 1995. One of the reasons for this belated development is the fact that an understanding for the potential value of the elucidation of the human genome spread only slowly in Japan. This began to change in 1989, when the Science Council of Japan published its memorandum On the Promotion of the Human Genome Project (HGP) which contained an ambitious outline for Japan's participation in the project. In 1994, the importance of genetic research was stressed in even stronger terms by the Human Genome Committee in a second report on Urgent Tasks to be taken for the Human Genome Project. The committee:

> strongly [urged] the governmental agencies and . . . the private sectors, to increase their support of HGP in Japan . . . If this is not done, the infrastructure of the life sciences in Japan in the twenty-first century will lag far behind those of other countries and will suffer from substantial loss in . . . various areas.
>
> (Human Genome Committee 1994: 21)

Concerning gene therapy, however, a small band of Japanese researchers closely monitored the early activities in the US. Since 1985, MHW sponsored a series of small-scale projects as Special Research Activities under its Health Sciences Research Grants program in which a group of researchers worked on problems of gene therapy while tracking US activities (Nakagome 1990). In October 1991, MHW formed a sub-committee on gene therapy within its Council on Health Sciences (*Kōsei kagaku kaigi*). Besides monitoring the international development of gene therapy, the sub-committee was to draft a research plan, investigate the feasibility of gene therapy, and reflect on the social implications of this treatment. Thereafter, the ministry finally started specifically funding gene therapy research with a budget of 150 million Yen in fiscal year 1994 (Koseisho 1995: 264) and 230 million Yen in 1995.

In 1994, steps toward the first clinical application of gene therapy were undertaken in Japan. In August of that year, Hokkaido University applied for the approval of a gene therapy protocol for the immuno-deficiency syndrome ADA following a US-American example. The technology used in this trial had to be imported entirely from the US, which made cooperation with US authorities indispensable to judge the safety of the vector. However, since this cooperation proved to be difficult, the physicians of Hokkaido University could begin Japan's first gene therapy only in August 1995 (Hannay 1995: 188). In August 1997, the physicians concluded the trial and called it a success (*Nihon Keizai Shinbun*, 5 August 1997).

The pilot trial at Hokkaido has contributed to the development of Japan's gene therapy in several important ways. First, it has demonstrated that the possibility of successful treatment with gene therapy is high. Second, it has directed public attention to the therapy. MHW took great care to make the approval process and the therapy trial itself transparent in order to produce a favorable public impression. Third, the trial helped to clarify the application procedure for such experimental therapies (Nikkei Baioteku 1995: 297). MHW's lack of clear regulations was a major reason why Japanese researchers could not implement a single therapy protocol since the beginning of the Hokkaido trial. Although according to a study conducted in early 1995, fifty-one medical institutions in Japan had planned to implement gene therapy protocols in the future and fifteen had replied that they intended to do so by the end of 1996 (Zaidan hojin 1996: 1), by the end of 1997, Hokkaido University remained the only one that did. However, meanwhile MHW put a two-track approval system into place that is closely modeled on its US pendant. Particularly, the Guidelines for the Safety and Quality of Materials Used in Gene Therapy are meant to reduce dependency on the US approval system and to encourage industry to undertake R&D in this area, for example on vector technology (Nikkei Baioteku 1995: 21).

A list of Japanese Universities that are now advanced in planning gene therapy protocols – all of which are still based on American technology – is given in Table 10.2. Especially noteworthy is Kumamoto University's application for approval of gene therapy for AIDS. In preparing this trial, the physicians from Kumamoto closely cooperate with the pharmaceutical company Green Cross and its American partner Chiron-Viagene. This trial would be a novelty in that, first, the protocol is an *in vivo* therapy which is a treatment inside the human body and second, the potential commercial value of this procedure plays a very immediate role in this trial. However, in June 1997, the approval of the protocol which was scheduled to start just a few weeks later was suddenly put on hold by MHW because of suspected irregularities in the data submitted during the administrative screening process (Nikkei Baioteku 1997: 276).

Since the Japanese guidelines for gene therapy limit testing to moribund patients, all other advanced protocols address different cancer treatments. All of them make use of procedures that have been already established as fairly effective by foreign trials. Of interest is the clinical test planned by Nagoya

Table 10.2 Selected Japanese gene therapy projects (as at end of 1997)

University	*Private partner*	*Treated disease*	*State of progress*
Hokkaido	Genetic Therapy	ADA	Termed 'completed' August 1997
Kumamoto	Green Cross Chiron-Viagene	AIDS	Approval withdrawn in Summer 1997
Okayama	Introgen Therapeutics RPR GenCell	Lung cancer	Applied at ministry level
Tokyo	Somatix Therapy	Renal cancer	Applied at ministry level
Nagoya	Kaken Pharmaceutical	Brain tumor	Approved at university level in November 1997

Source: Nikkei Baioteku (1997: 278).

University because here, for the first time, a vector would be used that is produced domestically by Kaken Pharmaceutical (Nikkei Baioteku 1997: 276).

'Japan must catch up': pharmaceutical industry and gene therapy

Already during the preparations for the implementation of Japan's first gene therapy protocol the opinion was voiced that greater independence from American help – especially in vector technology – should be achieved. Under the headline 'Japan must catch up,' *The Daily Yomiuri* (14 February 1995) called for an accelerated build-up of an institutional structure for gene therapy. *The Mainichi Daily News* demanded that at least organizations were to be put into place that would allow independent vector development and testing (9 February 1995). In addition, the Japanese ought to undertake more basic research because there would be a 'lot of room for breakthroughs.' This would allow Japan to establish an independent position in the field in spite of America's great lead.

However, as late as 1991, Japan's pharmaceutical industry showed only a limited interest in the implications of the human genome project (Matsubara 1991: 1546). Until 1994, Otsuka Pharmaceutical was the only Japanese drug manufacturer that undertook research on larger sections of human DNA. However, since then, the government has taken general measures and offered specific support to projects promoting genetically-based pharmaceutical technology. General measures include the reorganization of the Health Sciences Research Resource Bank (HSRRB) of the Health Sciences Foundation which houses a cell bank and a gene bank, and its opening to private companies. Support was lent to a string of specific projects involving industry: In March 1994, four pharmaceutical manufacturers founded the

Agene Research Institute which is exploring the genetic basis of aging, with the financial support of the Drug Organization (Kimura 1995: 93–4). In October 1994, the Kazusa DNA Research Institute was created with the aim to speed up Japan's human genome research by developing new research techniques (Swinbanks 1994: 6).

Already in 1995, MITI founded its own Biotechnology Center concentrating its resources on the elucidation of the genome of thermophile bacteria (Suzuki 1996: 2144–8). Also in 1995, MITI announced its Helix project which will develop new technology for genome research. Seventy percent of its cost will be carried by the Japan Key Technology Center which is administrated jointly by MITI and the Ministry of Posts and Telecommunications (MPT). Fifteen companies will contribute the remainder of the funds, among them the drug manufacturer Yamanouchi. The investors hope to accelerate the analysis and application of genetic functions by capitalizing on Japan's 'traditional' technological strengths, for example in semiconductors, robotics, and automatic manufacturing. Through this project, Japanese scientists are making a conscious effort to enter early into the second phase of genome research. In this phase, the elucidation of important functional genes whose patents are of particularly high value is at the center of activity. Besides the genetic sequences themselves, early patenting of the techniques applied in this process is also a declared aim of the project (Noguchi 1996: 1951–2).

MHW is supporting the activities of Japanese companies in the areas of vector research and finding therapeutically relevant genes as part of its Genox project. In March 1996, the Genox Institute was founded with the help of the Drug Organization. In this project which will study the genetic basis of allergic reactions, metabolic diseases of the brain, bone diseases and congenital heart diseases, participants do not only include pharmaceutical companies, but also apparatus manufacturers and other biotechnology oriented companies. The institute will promote basic research by preparing a cell bank and a cDNA-library.

More specifically for gene therapy, it was when the introduction of guidelines reduced regulatory insecurity that technological development became more attractive. In March 1995, the DNAVec Research Inc. at Tsukuba Science City was officially founded by seven companies and the Drug Organization. This project is hoped to alleviate Japan's dependency on American vectors which was felt so painfully during the preparation for the therapy trials at Hokkaido University. By the summer of 1997, first successes were reported and patents filed for an especially safe vector developed from a murine virus (Nihon Keizai Shinbun, 28 June 1997). However, there was still no Japanese vector system ready for clinical use and all protocols applied for MHW approval were dependent on US technology (Nihon Keizai Shinbun, 21 July 1997).

Also located in Tsukuba will be the facility for vector-testing which is under construction on the premises of the primate center of Japan's National Institute of Health (Nikkei Baioteku 1995: 297). This location will allow the facility to make use of the National Institute's primate colony for the tests. It is being established with the aim to promote the development of vectors and other tools for gene therapy by private companies (Zaidan hojin 1996: 1). However, by the end of 1997, there was still no Japanese institution able to test vector safety conclusively (*Nihon Keizai Shinbun*, 21 July 1997).

In spite of all these activities, the official Japanese response to the development in genome research in general and gene therapy in particular has been slow. Many Japanese companies have therefore decided to go their own, independent, way. In 1995, the director of Otsuka Pharmaceutical's leading-edge research into human cDNA criticized in an interview the inflexible stance of the Japanese government: The projects of MHW and MITI that are finally starting should have begun already in 1992; the turf-fighting of the ministries had delayed the proceeding. Otsuka would therefore not participate in these projects (*Nikkei Sangyo Shinbun*, 27 October 1995).

Chugai Pharmaceutical and Takeda Chemical Industries, also, are staying away from government-sponsored projects while pursuing an alliance strategy. Chugai has a history of a late, but successful entry into the application of genetically-based biotechnology to the development of pharmaceuticals. In 1984, it entered an alliance with the American company Genetics Institute which led to the development of the blood production stimulant Erythropoetin. Later, another cooperation with The Institute for Medical Sciences of Tokyo University led to the marketing of G-CSF (Yamaguchi and Shirubaman 1992: 97). Since 1995, Chugai is also trying to participate directly in the fruits of research into the human genome by setting up an Institute for Molecular Medicine and pursuing partnerships with American and Australian companies.

Takeda is following an alliance strategy, too. In March 1995, the company entered a cooperation with SmithKline Beecham, the UK's second largest pharmaceutical manufacturer. In particular, this agreement will give the Japanese access to SmithKline's genome data base. Although information and personnel exchanges are planned, the development of new products is to remain separate. Since the cooperation began, there are about thirty scientists at Takeda working on the human genome (*Nihon Keizai Shinbun*, 25 July 1996).

However, the greatest success of a Japanese company to date regarding specifically gene therapy has been achieved by a small brewery. Takara Shuzo in cooperation with Indiana University developed a method for transporting genetic material into cells which enhances the success rate of this procedure from conventionally 5 percent to more than 70 percent in certain human cells. This method could reduce the cost of gene therapy to about a tenth of the conventional level. Therefore, in October 1995, when MHW established a study team that was to promote the introduction of gene therapy into Japan, Takara's technique was viewed as a strong candidate for intensive support. The substance developed by Takara has been on sale since May 1997, and by the end of that year, an additional eight US institutions were seeking approval from the Federal Drug Agency to begin clinical trials (Nikkei Baioteku 1997: 276; *Nihon Keizai Shinbun*, 30 October 1997).

In summary, Japan's participation in genetic research and one of its central applications, gene therapy, which is among the areas of basic research perhaps most important to the future of the pharmaceutical industry, has been quite late. For example, DNAVec was only established at a time when there were already about twenty US firms developing vectors. With a view to the large American investment, public and private, in the field of genetic research it remains to be seen if the above projects which make use of traditional policy tools like the creation of research consortia, the provision of capital and some support services, will substantially strengthen the position of Japanese companies in this critical field of research.

R&D policy for the drugs of the twenty-first century?

Internationalization and cost-control measures confront Japan's pharmaceutical industry with increasingly harsher conditions in its domestic market. To ensure their long-term survival, companies will have to develop products that can be marketed internationally. While they must not only expand their R&D efforts, but also face up to the challenges of the 'biological revolution' and adjust their research strategies accordingly, Japanese companies have to struggle with a comparatively limited sales basis and deal with a domestic health care system that provides little incentive for risky R&D efforts. Japan's research promotion for the pharmaceutical industry draws strongly on traditional tools. While this may help companies to some degree with the creation of original technology, some of MHW's policies may have actually discouraged the creation of innovative drugs, for example the price reduction mechanism. Therefore, stimulating creative pharmaceutical research seems not to have been given a very high priority on the agenda of MHW (Anegawa 1997: 35). The recent dissolution of the Bureau for Pharmaceutical Affairs (July 1997) might indicate that after the AIDS scandal of 1996 and other problems (Nikkei Baioteku 1997: 174), the promotion of the pharmaceutical industry has become even less of a priority for the ministry.

What does the twenty-first century hold for Japan's pharmaceutical industry? Recently, the mergers and acquisitions fever in the global pharmaceutical market place is being viewed more critically. While some still view 'big investors' as the only major players of the future, others predict an increasing disintegration of the industry caused by the enormous complexity of the pharmaceutical process. A 'new division of labor in innovation' made possible by the 'biological revolution' might give new opportunities to medium-sized companies who so far 'lacked the size to found their competitiveness on innovation' (Della Valle and Gambardella 1993: 300–1). In this environment, alliances will be the key success factors. To establish such alliances, companies must have (1) the ability to assess the technological capability of a potential partner company; (2) a risk-spread of research performed in-house; and (3) successful basic research to produce original technology that can serve as a basis for alliances (Yamaguchi and Shirubaman 1992: 98–9). Companies will have to 'found their strategies on superior research skills and abilities in well defined niches, possibly by taking advantage of systematic linkages with their domestic scientific communities.' All this should encourage more openness of research, which is 'necessary to establish effective linkages with the international scientific community (Della Valle and Gambardella 1993: 301). Chugai's past successes provide an example of the successful application of the strategy outlined above.

Under this 'new division of labor in innovation,' some research promotion measures may even turn out to be hindering companies' R&D efforts. Research consortia which are encouraged by the Drug Organization may possibly prevent companies from forming more fruitful international alliances. This may be even more the case if consortia interfere with the free flow of information that is crucial to pursuing successful alliances. Thus the 'biological revolution' may render efforts to support Japan's pharmaceutical industry's R&D through traditional measures ineffective. On the part of companies, moves not to participate in government sponsored consortia and to pursue independent collaborations, for example with foreign universities, are spreading. In 1994, Japanese companies sponsored nineteen alliances with universities worth $319 million (*Nature Biotechnology*, 14 April 1996).

In this environment, policy may have to limit itself to providing a favorable environment for life sciences research. Several observers have emphasized the importance of expanding basic life science research, creating a national life science agenda for Japan, concentrating administrative oversight of the life sciences in one hand, strengthening the role of universities in the commercial exploitation of new technologies, etc. (Anegawa 1997, Hiroi 1997). For example, the number of life science research positions could be increased to reach a 'critical mass.' In 1991, there were around 37,500 bio-scientists working in the US, compared to only 8,500 in Japan.

In May 1997, MHW has reacted by promoting its Council on Health Sciences to the status of an advisory council (*shingikai*). Further steps to improve the situation are expected from the Basic Plan for Research and Development on Life Sciences which was developed by the Council for Science and Technology under the Prime Minister and officially adopted on 13 August 1997. The Plan states that 'intellectual property rights to DNA, proteins and other biological molecules . . . promise foundations on which the next generation industries will rise.' It further calls for 'placing priority on specific fields,' emphasizing areas with possible medical applications. Besides calling for an overall expansion of the life science budget which is relatively smaller than that of other industrially developed nations, the Plan lists specific measures to be taken (Prime Minister of Japan 1997). Taking into account its severe fiscal situation, Japan may not see a revolution, but at least a turn-around in the life sciences at the beginning of the twenty-first century.

In this improved environment, it seems still possible for Japanese businesses to enter the field of research on the human genome, especially as this research is moving into its second stage, focusing on the functions of genes which can be patented more readily. Concerning gene therapy, the call of the US National Institutes of Health for the solution of fundamental problems and less clinical trials (Orkin and Motulsky 1995) may also be a hint that Japan could still enter this area of scientific work successfully. The Genome Frontier Program, a 4 billion Yen project supported by the Science and Technology Agency (STA) and the Institute of Physical and Chemical Research (RIKEN), which is meant to help fill the gap between Japan and other industrialized countries in the field of human genome research, might provide some support from the public sphere. However, at this crucial moment, efforts are delayed by a debate if this project is to be limited to gene sequencing only or if functional elucidation which is necessary for winning patents should be undertaken as well (Saegusa 1997: 772). Japan's catch-up effort seems to last well into the twenty-first century, at least in the field of the life sciences.

REFERENCES

Anderson, W. French (1995) Gene Therapy. In *Scientific American*, September, pp. 124–8.

Anegawa, Tomofumi (1997) Seiyaku sangyō no R&D seisaku no keizaiteki bunseki (Economic analysis of pharmaceutical R&D policy). In *Iryō to Shakai* (Medical Care and Society), vol. 7, no. 2, pp. 1–35.

Della Valle, Francesco and Alfonso Gambardella (1993) 'Biological' Revolution and Strategies for Innovation in Pharmaceutical Companies. In *R&D Management*, vol. 23, no. 4, pp. 287–302.

Endo, Hisao and Nobuaki Tanaka (1997) Wagakuni no iyakuhin sangyō no kokusai kyōsōryoku no genjo to kanōsei (Japanese pharmaceutical industry's development capabilities and the potential to acquire international competitiveness). In *Iryō to Shakai*, vol. 7, no. 1, pp. 46–71.

Hannay, Timo (1995) Green Light for Gene Therapy in Japan. In *Nature Medicine*, vol. 1, no. 3, p. 188.

Hiroi, Yoshinori (1997) Kōteki seisaku to igaku, seimei kagaku kenkyū kaihatsu (Public policy and biomedical research). In *Iryō to Shakai*, vol. 7, no. 2, pp. 36–52.

Howells, Jeremy and Ian Neary (1995) *Intervention and Technological Innovation. Government and the Pharmaceutical Industry in the UK and Japan*. Houndsmill and London: Macmillan.

Human Genome Committee, Panel on Life Sciences. Prime Minister's Council for Science and Technology (1994), *Urgent Tasks to be taken for the Human Genome Project*. Tokyo: Unpublished.

Human Science (1995) June, pp. 31, 67, 69.

—— (1997) January, pp. 32–3.

Kimura, Bunji, Akira Fukami, Shin'ichiro Yanagisawa and Kumi Sato (1993) The Current State and Problems of Japan's Pharmaceutical Market. In Okimoto, Daniel and Aki Yoshikawa (eds), *Japan's Health System. Efficiency and Effectiveness in Universal Care*. New York: Faulkner & Gray.

Kimura, Hitoshi (1995) Eijīn kenkyūjo (Agene Research Institute). In *Tanpakushitsu kakusha kōso*, vol. 40, no. 1, pp. 93–5.

Koseisho (1995) *Kōsei hakusho (Heisei 7nenban)* (White Book on Health and Welfare (1995 Edition)), Tokyo: Gyosei.

Koseisho Kosei Kagakka (1993) *Kōsei kagaku yōran '93* (Overview of Health Sciences 1993). Tokyo: Kosei Kagaku Kenkyusho.

Koseisho Yakumukyoku (1993) *21seki no iyakuhin no arikata ni kan suru kondankai* (Forum for the pharmaceuticals of the twenty-first century), Tokyo: Gyosei.

Marshall, Eliot (1995) Gene Therapy's Growing Pain. In *Science*, vol. 269, pp. 1050–5.

Matsubara, Ken'ichi (1991) Hitogenomu kaisetsu keikaku no shinten to Nihon ni okeru purojekuto ni tsuite (On the human genome plan and its Japanese pendant). In *Tanpakushitsu kakusha kōso*, vol. 36, no. 8, pp. 1542–9.

Mitchell, Will, Thomas Roehl and John Creighton Campbell (1996) Sales, R&D, and Profitability in the Japanese Pharmaceutical Industry, 1981–1992. In Naoki Ikegami and John Creighton Campbell (eds), *Containing Health Care Costs in Japan*. Ann Arbor: University of Michigan Press, pp. 132–42.

Nakagome, Yasuo (1990) *Kōsei kagaku kenkyū hojokin (tokubetsu kenkyū jigyō) shōwa 63 – heisei gannendo identeki kumikae ni yoru idenshi chiryō kenkyū hōkokusho* (Health sciences grant (special research activity) 1988–1989 report on gene therapy based on genetic engineering). Tokyo: unpublished.

Nambu, Tsuruhiko (1997) *Iyakuhin no sangyō soshiki: Yakka kisei no keizaiteki kōka* (The industrial organization of pharmaceuticals: the consequences of price regulation in Japan). In *Iryō to Shakai*, vol. 7, no. 1, pp. 3–15.

Nature Biotechnology (14 April 1996), p. 442.

Nature Genetics (1996) Capitalizing on the Genome, vol. 13, no. 1, pp. 1–5.

Nihon Keizai Shinbun (25 July 1996), p. 41.

—— (28 June 1997), p. 14.

—— (21 July 1997), p. 35.

—— (5 August 1997), p. 34.

—— (22 September 1997), p. 42.

—— (26 October 1997), p. 1.

—— (30 October 1997), p. 5.

Nikkei Baioteku (ed.) (1995) *Nikkei baio nenkan 96* (Nikkei Bio-Yearbook '96). Tokyo: Nikkei BP-sha.

—— (1997) *Nikkei baio nenkan 98* (Nikkei Bio-Yearbook '98). Tokyo: Nikkei BP-sha.

Nikkei Sangyo Shinbun (27 October 1995), p. 21.

Nikkei Weekly (3 June 1996), p. 9.

Noguchi, Teruhisa (1996) Shinsangyō ni kitai sareru jisedai genomu kenkyū suishin no kiso gijutsu kaihatsu (Development of basic technology for the promotion of next-generation genome research hoped to become a new industry). In *Tanpakushitsu kakusha kōso*, vol. 41, no. 13, pp. 1951–2.

Orkin, Stuart H. and Arno G. Motulsky (1995) *Report and Recommendations of the Panel to Assess the NIH Investment in Research on Gene Therapy*. Bethesda, MD.

Prime Minister of Japan (1997) *Basic Plan for Research and Development on Life Sciences*. Tokyo: STA Homepage (www.sta.go.jp/life/lifeplan.htm).

Reich, Michael R. (1990) Why the Japanese don't Export more Pharmaceuticals: Health Policy as Industrial Policy. In *California Management Review*, vol. 32 (Winter), pp. 124–50.

Saegusa, Asako (1997) Genome Research Strategy Splits Japanese Scientists. In *Nature*, vol. 389 (23 October), p. 772.

Sapienza, Alice M. (1993) Assessing the R&D Capability of the Japanese Pharmaceutical Industry. In *R&D Management*, vol. 23, no. 1, pp. 3–16.

Shūkan Shakai Hoshō (1996) Yakuzai minaoshi no shisaku menyū wo rongi (Considering the choice of measures concerning drug-related costs), no. 1910 (21 October), pp. 26–9.

Somucho Tokeikyoku (ed.) (1996) *Heisei 8-nen kagaku gijutsu kenkyū chōsa hōkoku* (Report on the Survey of Research and Development 1996). Tokyo: Nihon Tokei Kyokai.

Suzuki, Osamu (1996) Tsūsanshō seihin hyōka gijutsu sentā baiotekunorojī sentā (The biotechnology center of MITI's center for product evaluation technology). In *Tanpakushitsu kakusan kōso*, vol. 41, no. 14, pp. 2144–8.

Swinbanks, David (1994) Institute to Boost Japan Sequencing Efforts. In *Nature*, vol. 372, p. 6.

The Daily Yomiuri (14 February 1995), p. 7.

The Mainichi Daily News (9 February 1995), p. 2.

Yakuji handobukku '96 (1996) (Pharma Handbook '96), Tokyo: Yakugyo Jihosha.

Yakuji handobukku '97 (1997) (Pharma Handbook '97), Tokyo: Yakugyo Jihosha.

Yamaguchi, Hidemaru and J.B. Shirubaman (1992) Nichibei baio iyakuhin jigyō no tenkai (Development of American and Japanese bio-pharmaceutical business). In *Zaikai kansoku*, vol. 57, no. 3, pp. 62–99.

Yoshikawa, Aki (1989) The Other Drug War: US-Japan Trade in Pharmaceuticals. In *California Management Review*, vol. 31 (Winter), pp. 76–90.

Zaidan Hojin, Hyuman Saiensu Shinko Zaidan (1996) Chōsa hōkokusho – idenshi chiryō ni kansuru kokunai iryō kanren kigyō no ninshiki (Report on the survey of domestic health care companies' awareness of gene therapy). In *HS Repōto*, no. 26, Tokyo: unpublished.

11

R&D CONSORTIA IN THE 1990s

National competitiveness and international cooperation in the case of semiconductors

Koki Inoue[1]

Introduction

Japan is adapting to the new information age of the post-cold war era by not only strengthening its economy, but also actively building up new international relations. This is also true in the area of Science and Technology (S&T) policy. The S&T Basic Law (1995) and the Cabinet's S&T Basic Plan (1996), both novelties enacted half a century after the surrender of the Japanese Empire at the end of the Second World War, are a kind of Japanese Manifesto for such change. Emanating from a feeling of crisis among the Japanese, both the Law and the Plan aim at heightening Japan's national competitiveness by strengthening its domestic R&D base and thereby increasing Japan's contribution to the world. There are certain S&T policy areas such as the liberalization of government regulations to improve the environment of research and development (R&D) activities in Japan which would lead to both increased Japanese competitiveness and contributions to world S&T efforts. However, in the area of governmental R&D project policy, there seems to be a 'salad bowl' consisting of certain projects mainly for competitiveness and certain projects for international contribution. Nonetheless, under the expected budget constraints of the twenty-first century, there is pressure to establish a new type of 'dual purpose' R&D project policy where both purposes are sought in a single project.

This paper examines the two characteristics appearing in both the S&T Law and the Plan: governmental S&T policies for 'national competitiveness' and for 'international (global) contributions.' Herein I present the case of the semiconductor industry which is one of the most critical areas of S&T policy.

Resource needs for semiconductor R&D tend to surpass individual companies' resources even for manufacturing technology, and at the same time, accessibility and uncertainty of manufacturing technology have increased markedly. Furthermore, the Japanese semiconductor industry has become increasingly 'sandwiched' between its US and Asian competitors as seen in the following section. Thus it has become more difficult for Japanese electronics companies to carry out all the activities for advanced semiconductor R&D internally or even within company groups. They have started redefining their competitiveness and begun restructuring their R&D Test and Evaluation organizations and activities, and have set up a private consortium for R&D Test and Evaluation. The Japanese government, namely the Ministry of International Trade and Industry (MITI) has also begun a national R&D project, called the Super Advanced Electronics Project, mainly for semiconductor manufacturing technology, ending the twenty-year 'wait-and-see' policy after its Very Large Scale Integration (VLSI) project. The private consortium and the government-sponsored one are natural and necessary responses to the industrial-technological situation, as well as one to the problem of 'sandwiching.' They are necessary for maintaining national excellence and competitiveness, and in this sense seem to fit the post cold-war era of increased global competition.

However, there remains the issue whether MITI's new project's aims, activities, and character should be changed from only national competitiveness to international contribution through cooperation among R&D projects (or consortia) of various nations. MITI's ASET (Association of Super-advanced Electronics Technologies) project, the US DARPA (Defence Advanced Research Projects Agency) project, and the European MEDEA (Micro-Electronics Development for European Applications) are all exclusively aimed at assisting each country's semiconductor industry to maintain its technological level, countervailing effects of increasing industrial dominance, and – as a side-effect – they all contribute independently to heightening the world's common level of semiconductor manufacturing technology. The increasing accessibility of manufacturing equipment or other forms of manufacturing technology also serves to raise the world's level of technological achievement. Semiconductor industrial-technological development on a world scale is critical for sustaining the development and welfare of world society. Therefore, MITI's ASET project, the US DARPA project, and the European MEDEA seem able to achieve better results and save R&D resources by coordination of parts of each project. In the remaining part of each project, they may be able to seek national competitiveness, although the degree to which projects affect competitiveness appears to be decreasing. Because of this, a part of the government-sponsored 'national' R&D projects, particularly those industry-oriented, should be internationally coordinated.

Japanese chip makers' R&D restructuring

It has become critical for the Japanese semiconductor industry to restructure R&D activities. In general, this is true for every large chip maker of any regional ownership for the following two reasons. First, in the semiconductor industry, the more advanced the generation of a chip is, the more R&D resources (facilities, funding, personnel, etc.) are required for its development, and the growth rate of the latter tends to exceed the increased pace of industrial sales. It becomes more and more difficult for chip makers to manage R&D resources with every chip generation.[2] Today, it is said that some ten billion dollars are necessary for just device and process (a part of manufacturing) technology development for the 256 mega (M) bit DRAM (Dynamic Random Access Memory) or 0.25μm technology (the narrowest line width for chip processing). Needless to say, chip makers have to invest also in factories and operations, as well as R&D activities other than manufacturing technology, mainly being for chip architecture. Capital is not the sole issue, but human resources, facilities, and information are all problems for manufacturing technology R&D, where the scope of necessary technologies has been widened vastly. In the future, there may be even some peculiar periods with leaping – not continuous – expansion of R&D needs, i.e. for the 1–4 giga (G) bit DRAM or 0.18–0.1μm technology.

Second, accessibility of technological information to outsiders (or 'free riders'), and the uncertainty or variability toward a technological solution, have both been increasing. Semiconductor manufacturing equipment is a good example. The more the equipment embodies fine processing technologies, the more critical it is for chip makers' advanced mass production (and ultimately their profits and growth). But once the processing technology is embodied in equipment as merchandise of equipment suppliers, it will be easily accessible to any chip maker at home or abroad, of any country's ownership, without asking whether the chip makers helped with its development or not.[3] Thus, through the body of equipment, valuable technological knowledge of the chip maker who used his own R&D resources and helped with the development of equipment will spill over and give so-called consumers' benefit to the outside chip makers (or 'free riders') who thereby could save on their own R&D expenditures.

In the real world, there are few 'free riders' worthy of the name, but there are many 'follower' chip makers slightly lagging behind the leader. They are able to quickly catch up with the top makers and save R&D resources over the advanced makers. Of course, advanced makers can achieve larger earnings through advanced cost reduction, preferential occupancy of the market for the most advanced chips at a rather high price, or collection of fees for the consumption of their intellectual property. Here are opportunity costs for both followers and leaders, and there can be a kind of equilibrium where both can be satisfied with their position from top to bottom according to their resources. However, if accessibility of technological information (for example through equipment) to the followers increases, then more makers will move to lower positions from an absolute technological view. Top makers will also fall back slightly and the bandwidth of players will become narrower. The whole technological pace of progress will become slower comparatively because the advanced makers' earnings will decrease and the followers' savings of R&D resources will increase by free or cheap information, if other conditions remain the same.

On the other hand, an increase in uncertainty or variability of technological solutions at an absolute level of technology leads to an increase in R&D resources required; it leads everyone (including the top maker) to give up developing their own solutions; and it creates a similar situation to the case of the above 'free riders,' because the individual value of information will be decreased by its uncertainty. In both of the above cases, one would expect to see an overall satisfaction in lower technological positions and, with a view to the absolute technological level, a comparably slower pace of progress. This is the rationale for government to intervene and sustain the tempo of semiconductor development.

The business situation that urges Japanese chip makers to restructure R&D, and that wherein Japanese makers have to do this, is a 'sandwiched' one between the US microprocessor and Asian DRAM manufacturers. Although this is not to say that only the Japanese semiconductor industry is in a difficult position, it appears that the Japanese semiconductor industry stands at a crossroads. The following factors have made the Japanese situation 'sandwiched':

1 the Korean catch-up in DRAM manufacturing;
2 rapidly increased demand (in value) in microprocessors for personal computers (PC) rather than in other types of ASIC (Application Specific Integrated Circuit) mainly for audio-visual apparatus;
3 US advantages and corporate strategies in microprocessors or digital logic chips for PC/networks; and
4 US efforts to recover or maintain their leadership in manufacturing technology under SEMATECH or I300I.[4]

These factors combined create the 'sandwiched' situation as follows: First, the share of Japanese and US semiconductor companies in the world market for semiconductors is gradually decreasing compared to other Asian producers, while the shares of both Japanese and US companies roughly match each other. The Japanese companies' share has halted its increase and started declining in 1990, and fell slightly under 40 percent in 1995. Other Asian, mainly Korean, companies have surprisingly increased their production, and their market share reached 12 percent in 1995. Taiwanese semiconductor manufacturers are also planning to increase sales beginning in 1996–1997. US companies have restored their share since 1990 (company shares based on nationality of ownership according to Dataquest).

Second, Japan's domestic semiconductor market has been shrinking as a share of the world market. It is estimated at roughly 25 percent of the world market ($129 bio.) in 1996. The US has reversed its market share since 1993, which at present stands at around 32 percent. Again, the share of other Asian countries has been rapidly increasing and may reach 20 percent in 1996 (WSTS 1996 autumn). A switch between Japan and other Asian countries in market size may possibly happen in 1998–1999. Although the exact figure depends largely on foreign exchange rates, the basic trend seems quite stable.

Third, the US has the SEMATECH consortium and its subsidiary I300I. This is one of the causes of the above results, but at the same time, it is a very important factor in itself that leads Japanese companies and government to set up various Japanese consortia. It helps the involved Japanese parties to give these consortia their 'national' character.

The 'sandwiched' situation has the following implications. First, concerning product types, it has become critical for Japanese chip makers to learn how to produce various new types of chips other than DRAM or conventional micro-controller/ASIC for stand-alone audio-visual apparatus. Second, at the same time maintaining excellence in manufacturing technology is important because of the following points: DRAM remains important in view of world market size; keeping some share in the DRAM market is critical for sustaining companies. DRAM requires cutting-edge manufacturing technology even in the future. Manufacturing technology is also important for the development of various new types of chips as far as the concept of 'development' includes manufacturing, and as far as chip architecture requires information about affordability of manufacturing. Thus manufacturing technology and its R&D remains critical for Japanese makers, as well as business and technological efforts for new types of chips with new architecture.

Concerning geographical areas, 'the global market' has become critical for the first time under the conditions of the information age, especially in order to set world-wide standards through alliances. At the same time, the domestic market of Japan seems more critical for Japanese chip makers to re-develop with new types of chips because the Japanese market may work as the first test field for Japanese chip makers' new products. Of course the Japanese market should remain open to any company of foreign ownership. The more this will be the case, the more important new types of chips with new architecture will become for Japanese makers.

The need for R&D resources and for sharing uncertainty and accessibility of various technologies is pressing the leading semiconductor companies in the world to take various measures. This is the most important, basic factor for restructuring efforts. However, at the same time, the 'sandwiched situation' in particular hastens Japanese semiconductor companies' reorganization, including the set-up of private consortia.

Japanese semiconductor companies have been undertaking various reforms. First, the most important point of restructuring from the perspective of Japanese chip makers is neither technology nor R&D, but business. Through their business strategies they are aiming for international alliances in order to establish world-wide *de facto* or any other standards for various new types of chips rather than technology.

Second, they are undertaking re-engineering efforts ranging from R&D to sales. They endeavor to restructure their organization and human resource allocation, including R&D. For example, they have reduced the number of researchers in their corporate 'central laboratories' and have sent them to their semiconductor business divisions' laboratories and development centers (Inoue 1996).

Third, they reinforce their engineering abilities for chip architecture. This is not done only by simply increasing the number of chip architects in their companies, but by various transactions such as cooperation with academia (for example through STARC (Semiconductor Technology Academic Research Center), see below), corporate alliances among chip makers, subcontracting work to other companies, vertical alliances with chip-application apparatus makers, and so on.

Fourth, regarding manufacturing technology R&D, Japanese chip makers undertake various cost-cutting measures, as well as human resource reallocation, as seen above. But this is not enough. Today, roughly 46 percent of newly employed engineers in Japanese semiconductor companies are engaged in 'Devices and Process Technology R&D' and 'Manufacturing Technology (production, inspection, and analysis techniques and technologies)' (Nihon Denshi Kogyo Shinko Kyokai 1995: 91). Reallocation of researchers and engineers, like the one mentioned above, from central laboratories to units on the business line level along re-engineering lines does not lighten the R&D burden significantly. It is becoming too difficult for single Japanese semiconductor manufacturers to carry this burden by themselves, so that they cannot solve the problems within their company alone. Therefore, they set up various private consortia in order to overcome each company's limitation in manufacturing-related R&D Test and Evaluation.

Rebirth of R&D consortia – private and public

In 1996, all of Japan's large semiconductor companies together have formed a private 'national' Test and Evaluation consortium for the front-end processing equipment of the next generation: SELETE (Semiconductor Leading-Edge Technologies, Inc.) (see Table 11.1). Its aim is to avoid duplication of work and waste of time by each company. The participating companies seek investment efficiency for their R&D Test and Evaluation, and find it beneficial to join these activities. For a long time, they have competed rather than cooperated with one another for manufacturing technology, even in the famous 'VLSI project' of the 1970s. However, in the latter half of the 1990s, they developed the ability and the will to set up a consortium without any help

Table 11.1 Structure and activities of SELETE

Aims and activities	1 To evaluate the next-generation front-end processing equipment 2 To research and develop the pre-competitive, generic technologies for semiconductor manufacturing (for manufacturing technology CAD). Evaluation is to be done for so-called 'a-version' of the equipment suppliers about • a few focused basic performances of each equipment • productivity and reliability by the unit-process engineers assigned from the member companies (their semiconductor division's development dept. *plus* the corporate lab.).
Legal form	A private joint-stock company
Established	In February 1996 by the ten major Japanese chip manufacturers: NEC, Toshiba, Hitachi, Fujitsu, Mitsubishi Electric, Matsushita Electric Industrial, Sharp, SONY, Oki Electric Industry, Sanyo Electric
Stockholders' equity	5 billion Yen
Total investment 1996–2005 (excluding the equipment to be evaluated)	35 billion Yen
Planned duration	10 years
Manpower	Assigned engineers: around 85 persons in 1996 (300 persons planned in 1997)
Main facilities	A new cleanroom of 2,000 sq m (in the Hitachi Yokohama manufacturing Technology Laboratory area)

Source: Table compiled by the author based on information received from responsible people at the organization concerned during 1996. Copyright and any errors belong to the author.
Note: 'Next-generation' means the technology for 0.25µm-design rule and 300 mm diameter silicon wafer.

from government, contrary to the VLSI case. Such a private consortium for manufacturing-technology Test and Evaluation is surely epoch-making for Japanese semiconductor companies. It should be taken as a private 'national' consortium, distinct from corporate alliances between individual companies. And, correctly speaking, SELETE is more a 'Test and Evaluation' consortium rather than a R&D consortium.

The same Japanese semiconductor companies have also set up a private national center for industry-academia cooperation, STARC (see Table 11.2). In addition, the government is also promoting government-academia cooperation by enforcing competition in the proposal-selection of research grants for academics and by allowing more human resource exchange between academia, industry, and national laboratories.

The founding of SELETE and STARC were both prepared through the framework of SIRIJ (see Table 11.3), an industry's cooperative think tank through which Japanese semiconductor producers collectively formulate proposals to government and coordinate their policies.

Table 11.2 Structure and activities of STARC

Aims: Encouragment of	1 Universities' silicon semiconductor research activities 2 Education of graduate students for silicon semiconductors
Main activities	1 Management of research projects with universities for silicon semiconductors 2 Management of intellectual property out of its research projects through the: • presentation of industry-oriented needs for semiconductor research (mainly for silicon semiconductors) • reception of research proposals from university faculty researchers • workshops
Legal form	Joint-stock company
Established	December 1995 by the same members as SELETE
Stockholders' equity	100 million Yen (equal contribution by stockholders)
Annual budget	Around 100 million Yen
Average new entry number of projects to foster	Around 5 projects a year
Average duration of each project	Around 4 years

Source: Table compiled by the author based on information received from responsible people at the organization concerned during 1996. Copyright and any errors belong to the author.

Table 11.3 Structure and activities of SIRIJ

Aims and activities:	To develop a vision and an implementation plan for the semiconductor industry that would promote its position in Japan and in the world
Character of the organization:	A private think tank dedicated exclusively to the semiconductor industry and set up by the main members of the EIAJ (Electronics Industries Association of Japan) Electronics Device Department
Established	April 1994
Original members	NEC, Toshiba, Hitachi, Fujitsu, Mitsubishi Electric, Matsushita Electronics Industry (subsidiary of Matsushita Electric Industrial), Sharp, SONY, Oki Electric Industry, Sanyo Electric
Newly entering members	Rohm, Texas Instruments Japan, Motorola Japan

Source: Table compiled by the author based on information received from responsible people at the organization concerned during 1996. Copyright and any errors belong to the author.

In addition to the above, the government (MITI) has also begun a new R&D project named 'Super-Advanced Electronics Technologies Development Promotion Project,' which covers semiconductors, hard disks and drives, and flat panel display technology. Work on semiconductors is the largest and the most important part in the project. It is mainly aiming at producing elementary semiconductor technology for post-next-generation front-end processing. For this project, a new national R&D consortium called ASET has been created (See Table 11.4).

Almost twenty years have passed since the famous VLSI project was launched in 1976 until today's ASET was founded. Today's altered situation of the semiconductor industry is the most important and basic factor for MITI's project, as well as for Japanese industry's 'sandwiching' problem, the latter of which has helped MITI to make ASET into a 'national' undertaking. Particularly, the existence of SEMATECH has been helpful because it serves to keep the US government (including Congress) and industry silent against ASET. The reason why MITI has had no R&D project of 'national' scale concerning silicon semiconductor Large Scale Integration (LSI) between VLSI and ASET was fear of friction with the US on this issue. An exception was SORTEC, an R&D project for synchrotron X-ray lithography, but its scale was rather small and it played just the role of a pinch-hitter in the event although it attained certain results. This time, MITI can continue its project safely without having to be concerned about international frictions.

Thus Japan has at least three kinds of national consortia or organizations for cooperative R&D Test and Evaluation among government, industry and academia in the field of semiconductor technology: ASET, SELETE, and STARC, along with a policy think tank called SIRIJ (Semiconductor Industry Research Institute Japan). They all have a particular aim and activity area, so there is a 'division of labor' among them.

However, several issues remain. First, ASET is too technology oriented, especially for its X-ray lithography R&D. ASET's mission should have been more clearly geared toward R&D for 'adaptability' of technologies (i.e. its transferability to the IC factory floor), for choice of various technologies, and for introduction of attained technologies to result in factory production. But in the reality of ASET's early stage, this was not the case. One reason is that ASET mainly consists of researchers of corporate central laboratories who are technology-oriented while unconcerned with cost efficiency of technology. Since SELETE is a Test and Evaluation consortium, it consists of engineers of business divisions' laboratories and development centers. The latter are aware of the difficulties of the application of technology (embodied technology in equipment) at the factory level. The 'pure technology' bias of ASET was mainly the result of the failure of ASET's member companies to ask each central laboratory to plan ASET's technological objectives.

Moreover, there was MITI's confusion in controlling the project. Surely it is necessary for MITI to clearly differentiate ASET's area from SELETE's.

Table 11.4 Structure and activities of ASET

Structure	
Legal form	Industrial technology research association, approved by MITI, under the Act of Mining and Manufacturing Technology Research Association of 1961 (non-profit special corporation for cooperative technology research)
Aims	Promotion of research for the post-next-generation electronics technology which is generic, basic, and effective for the twenty-first century informaion industry and society (especially for the years after 2010). Semiconductor (silicon LSI), information storage, and flat panel display technologies are the most representative ones, and are viewed as 'technology infrastructure' for the advanced information society of the next century. It is regarded necessary and effective for government and industry to cooperate in R&D for the post-next-generation technology; therefore, government helps and coordinates research for such technology.
Established	February 1995
Approved by the Minister	April 1995
Planned duration of projects	Around 5 years

Member companies and project participation

Semiconductor = No. 1; Hard disk and drive = No. 2; Flat panel display = No. 3

Total 21 members	*Area 1 = 16 members*	*Area 2 =6 members*	*Area 3 = 8 members*
1 NEC	1	2	3
2 Toshiba	1	2	3
3 Hitachi	1	2	..
4 Fujitsu	1	2	..
5 Mitsubishi Electric	1	..	..
6 Matsushita Electric Industrial	1	2	..
7 Sharp	1	..	3
8 SONY	1	2	..
9 Oki Electric Industry	1	..	..
10 Sanyo Electric	1	..	..
11 NTT Advanced Technology	1	..	..
12 TI-Tsukuba R&D Center	1	..	..
13 Advantest	1	..	..
14 JEOL	1	..	..
15 SPC Electronics	1	..	..
16 Toshiba Machine	1	..	..
17 IBM Japan	..	..	3
18 Merck	..	..	3
19 Dainippon Ink and Chemicals	..	..	3
20 Japan Synthetic Rubber	..	..	3
21 Sumitomo Chemical	..	..	3

continued

Table 11.4 continued

Flow of funds and research organization

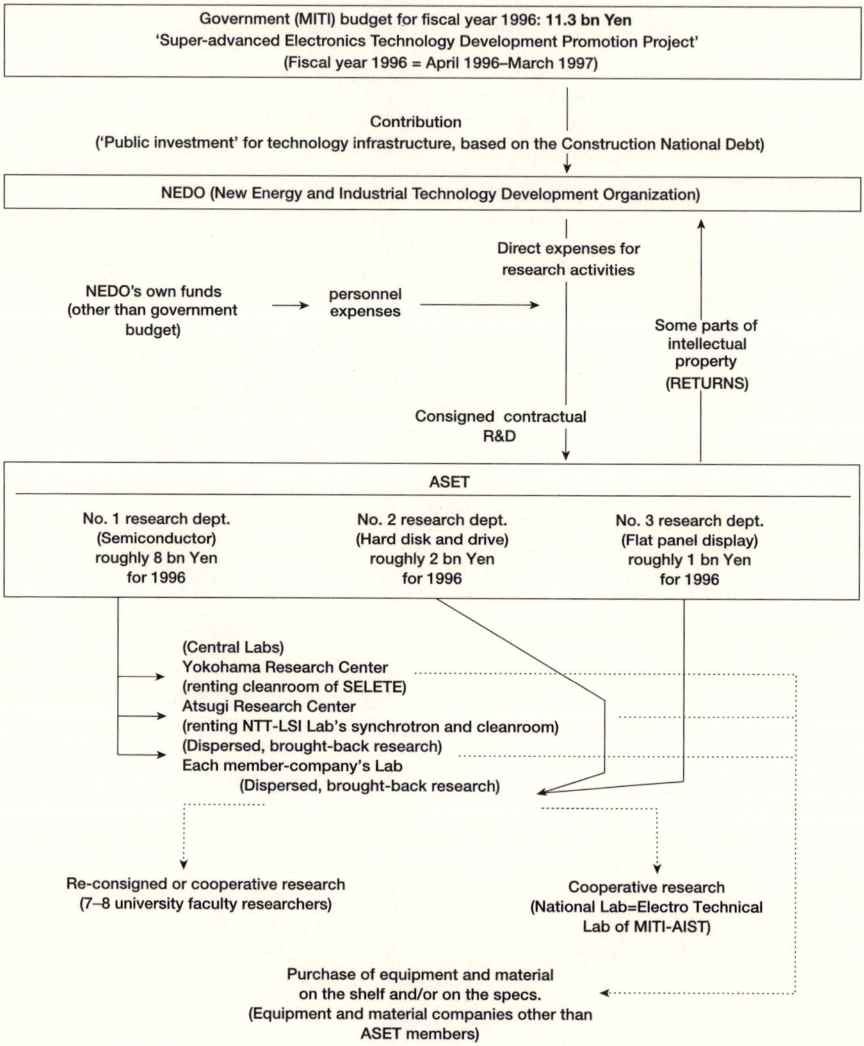

Table 11.4 continued

Technology, facilities and personnel

Technology areas:

Semiconductor = No. 1 research dept. (Technology for 0.18 μm-design rule)

1 Proximity X-ray Lithography technology, including the resist (Atsugi Center)
2 ArF Lithography technologies, including the resist (Yokohama Center)
3 Mask and mask-making equipment technologies for ArF and X-ray lithography (Atsugi and Yokohama Centers)

4 Electron Beam Lithography technologies	(brought-back
5 Plasma-process measurement, analysis, and control technologies	research in
6 Micro-contamination control and cleaning technologies	each company)

Hard Disk and Drive = No. 2 research dept. (mainly for GMR-related technology)
7 High density media (disk) technologies, including material design, thin-film processing, and structural analysis for magnetization
8 Device and their thin-film technologies (mainly for drive head), chiefly for Giant Magnet Resistance device

Flat Panel Display = No. 3 research dept.
9 Various electro-optics device technologies, including new type of LCD

Atsugi R&D Center of the No. 1 research dept. for proximity X-ray lithography roughly 90 researchers (scientists and engineers); from member companies 15 persons per company from 6 companies participating in this area:
NTT-Advanced Technology (originally from NTT itself), Fujitsu, NEC, Toshiba, Mitsubishi, etc.

Yokohama R&D Center of the No. 1 research dept. for ArF lithography (and plasma) roughly 40 researchers from top 10 chip manufacturing member companies.
Researchers from member companies are assigned on a part-time basis, not full-time, to the ASET projects.

Source: Table compiled by the author based on information received from responsible people at the organization concerned during 1996. Copyright and any errors belong to the author.

ASET should fit the government's S&T policy, which in turn should fit WTO rules prohibiting subsidies for industry. So MITI wanted to make its project as clearly 'R&D-oriented' as possible. Therefore it chose the post-next-generation area which means pre-competitive research. This is an end in itself, but the 'target generation' of technology and the 'R&D way' that leads to that technology are different. In view of the research 'way' to the desired technology, MITI should promote R&D that is also adaptable to real industrial usage, and show such intention in its public bidding papers clearly. However, this problem seems to be solved.

A second domestic issue, concerning the first problem, is the need to build up closer coordination between ASET and SELETE. ASET will bear some technological attainments and data in the future, but it should also be customer-oriented. In this case, the customers are IC factories. Since SELETE tests and evaluates technology embodied in equipment for IC factories, there should have been a close coordination *prior to* ASET's R&D activities. However, these two problems appear to be solved gradually, because the parties concerned have started to communicate better, and ASET's staff are now beginning to understand the problems.

A third difficult issue is whether Japan should draw up its own 'national technology roadmap,' like the US has done through the SIA (Semiconductor Industry Association 1994), SEMATECH (Semiconductor Manufacturing Technology Initiative), SRC (Semiconductor Research Corporation), as well as national laboratories and the Department of Defense (DoD). For such a roadmap, a reliable relationship between government and industry is a must, which means that the intentions and roles of both parties should match and work toward a clearly-defined national goal. In present-day Japan, this may be difficult to achieve.

Our most important issue is not a domestic one, but an international one. Before addressing it, it is necessary to consider the Japanese government's and industries' R&D situation. What is new in this semiconductor case? The most striking characteristics of the new consortia are represented by SELETE, not by ASET. SELETE is a Japanese semiconductor industry's *private* consortium, where competing companies join some of their funds and cooperate in some activities voluntarily. In Japan, this was unimaginable in the past, but the economic and technological situation of Japan's semiconductor industry, as described above, has pressed them to set up SELETE.

What is the difference between SELETE and the past VLSI project? The aim of VLSI was to catch up, while SELETE's aspiration is to maintain the technological level already reached, although there is no formal declaration of this goal. This is due to the different position of Japan's semiconductor industry in the world, as well as the different world situation. Surely Japan's semiconductor industry is 'sandwiched' today, but it is not as weak as it was at the time of the VLSI project. It seems certain that the most important factor to set up ASET and SELETE is the industrial-technological need for increased R&D resources and for sharing uncertainty and accessibility of various technologies, rather than the 'sandwiched' situation. But the 'sandwich' problem, as well as the existence of SEMATECH, gives SELETE and ASET their 'national character.' Although there is not much difference in the point of 'national' character between VLSI and ASET/SELETE, the underlying factors are very different between them. However, while the aims of VLSI and SELETE are different, regarding their activities, they are very similar.

Finally, there are some fundamental differences between the past VLSI project and today's ASET concerning the power relationship among the parties concerned. In the VLSI project, MITI had 'some' leadership before liberalization of trade and direct investment for large scale computers and LSI, although surely industry asked government to undertake the project. MITI had a clear objective to promote the industry composed of companies with Japanese ownership. So the VLSI project matched what could be called an 'industrial policy,' especially for specific industries included according to MITI's frame of reference. However, today, MITI does not have such 'leadership,' nor power or intention to pursue an 'industrial policy' for a specific industry. On the other hand, given the politico-economic balance of power in Japan, semiconductor manufacturers or any other specific industry as a single entity are not strong enough to have MITI or the government decide a project or a policy in their favor. Today's project is built on a vague understanding among all parties such as the semiconductor industry, MITI, and the Hashimoto Cabinet that innovation in information technology is important and semiconductor industry is at its basis. In this politico-economic process, it is rather the stagnation of the Japanese economy and R&D expenditure, as well as the new S&T policy than the specific situation of the semiconductor industry that has brought about the common understanding among the parties involved. This change in the political environment has occurred throughout the 1980s. But particularly after the collapse of the political stability formerly provided by the ruling Liberal Democratic Party (LDP), it has become more difficult for MITI to show leadership with specific individual 'industrial policies' against the resistance of a group of very large companies. Of course this does not mean the complete wane of MITI's influence over or through the administration (i.e. the cabinet). MITI, or more correctly MITI officials, seem unable or unwilling to 'lead' the Japanese semiconductor industry, but they hope to help industry, companies, and divisions 'coordinate' with each other for directions that are advantageous for innovation in information technology. This attitude appears to be similar to that of the US government's position in the world of the post-cold war era.

Following economic stagnation, Japanese R&D expenditure has been stagnating throughout the first half of the 1990s. This R&D stagnation is the first since the end of the Second World War, since R&D statistics were computed. This arouses a feeling of national crisis among industry, government (ministries, Cabinet, the National Diet), and academia. It has brought about the S&T Basic Law (1995) and the cabinet's formal decision on the S&T Basic Plan (1996).[5]

R&D expenditure in Japan has been declining for two consecutive years since 1993, because the decrease of private industries' spending was larger than the increase in the government's. Notably, Japanese industries' R&D expenditure is not too small in view of the ratio to GDP when comparing it to other industrialized countries; but the government's expenditure is. The US government spends much larger funds on R&D even as a percentage of GDP. Certainly this is nothing new, and US government figures include military R&D. But since around the end of the cold war, military R&D has been becoming more and more 'dual-use' oriented, and its success appears to be mainly in the information technology area including the Internet, semiconductors, etc. Therefore, Japanese governmental promotion for R&D has to become more strategic as well.

Economic stagnation, especially the stagnating situation of the Japanese information technology industries like computers, AV electronics, communication equipment, semiconductors, software, etc. is at the root of the feeling of national crisis, and the stagnation of R&D expenditure adds to this. Both led to the creation of the S&T Basic Law. This bill, which was sponsored by a cross-party group of Diet members, has passed the National Diet with almost unanimous approval in spite of political turmoil, representing such a feeling of crisis.

As for the national R&D 'system,' many elements in the system that have long supported Japan's development are turning out as hindrances, and tend to lack flexibility and competitiveness, and instead hamper cooperation and exchanges beyond organizations, says the S&T Basic Plan. The Plan articulates the need for construction of a new R&D system, especially for a system of cooperation and personnel exchange among private industry, universities, and national research institutes. Measures taken include the limited-term employment system and the smooth permission of side-jobs for researchers in national research institutes and universities, as well as the expansion of competitive open bidding for research funds by government and industry.

However, as mentioned in the introduction to this paper, there are such S&T policy areas like the R&D project policy where there exists just a 'salad bowl' of projects, some mainly for competitiveness and others mainly for international contribution. And MITI's Super Advanced Electronics Technology R&D Promotion project (or ASET consortium) belongs to the former group.

The above facts give a broad background of the conditions for achieving the budget under which the Super-Advanced Electronics project was established by MITI, with support from the semiconductor (electronics) industry and the Hashimoto Cabinet, as one of the measures of the 'Promotion plan of highly advanced information technology in Japan.' The Super-Advanced Electronics project is essentially meant to enhance national competitiveness, but this agenda must be changed in order to seek national competitiveness *and* international coordination.

The case for international coordination

In any possible international cooperation, one has to think about the US situation. The US government and Defense Advanced Research Project Agency (DARPA) seemed to change their Dual Use Technology (DUT) policy around 1994/95. In the narrow sense, DARPA's DUT program is becoming more and more selective and limited. The US DoD's direct subsidy to

SEMATECH has ended, following SEMATECH board's proposal. US major semiconductor makers such as Intel preferred to end such subsidies because of SEMATECH's success. DARPA and the US Congress welcomed the industry's decision. Of course, DARPA can and will go on with its 'own' R&D project for X-ray lithography, for example, in the name of national security. And the whole US government as well as DoD is promoting DUT in the broad meaning, especially through promotion for various international standardization issues ranging from the Internet to encryption technology. Also the US government-industry division of labor continues, like the US Semiconductor National Technology Roadmap. Therefore, Japan reacted by setting up 'national' consortia.

However, SEMATECH itself has become a US national 'private' R&D consortium. Thus it can move freely as a private entity without congressional approval. On the other hand, some part of SEMATECH's funds will be removed, and SEMATECH or its member companies have to reconsider the role and activity of SEMATECH under such conditions. One outcome of this is SEMATECH's subsidiary for the International 300mm Initiative (I300I) joint venture, which matches Japan's SELETE (and J300[6]).

Japan should continue to sponsor consortia, but such Japanese consortia should advance various kinds of international coordination. There are several coordination cases already existing. First, there is 'international coordination' in the area of 'standardization' concerning 300mm wafer and such wafer-oriented semiconductor manufacturing equipment. SIA and EIAJ (Electronics Industries Association of Japan) have already agreed to promote international coordination concerning standardization, as one of the activities of the (World) Semiconductor Council, which was affected by the discussion about the termination of the Japan–US Semiconductor Agreement. This industry-based framework has been widened to include the EU and Korea. Although the contents of the coordination matters are not yet clear, especially concerning whether it includes the 300mm wafer and related equipment standardization, maybe such a framework will help further activities for that standardization. Actually I300I and J300 have started coordination for such 'standardization.'

Second, there are also international coordination activities between SELETE and I300I about the 'evaluation' of the next-generation equipment. This 'international coordination' takes place between a 'national' consortium (SELETE) and a national consortium's international subsidiary (I300I). This is possible because, although SELETE is surely a 'national' consortium in view of its stockholders, it is also an 'internationally open' venture with regard to the conditions of consignment to it. Any foreign chip manufacturer can consign under certain conditions (as a customer of SELETE). In this respect, there is little difference between SEMATECH's subsidiary I300I and SELETE.

Now it seems necessary for the national R&D consortia to find new ways of international coordination. There is an unsolved important issue for international coordination among national R&D consortia, such as between ASET and SEMATECH (or I300I). The most important basic factor that has led to the creation of ASET (and SELETE) may also promote such coordination. That is the industry's need for R&D resources and for sharing uncertainty and accessibility of various manufacturing technologies, especially concerning advanced lithography technology.

There is already some framework for international exchange of technological attainments and activities involving Ultra-violet eximar LASER lithography including SEMATECH (SRC), Japanese parties belonging to ASET, and European parties belonging to MEDEA, by the form of international convention. This framework needs to be expanded.

Up to the present, manufacturing technology has directly affected corporate profitability. However, today this view is becoming too vague. As seen above, for example of lithography, there are many sub-technologies for each technology. Needs and risks are becoming too large even for national consortia, which have many other tasks than lithography. In short, today some area of manufacturing technology R&D is turning from a profitability problem into one of affordability and adaptability for choices and real introduction of technologies, comparing the schedules and constrained resources. In practice, the most critical point will be to get the decisions of many leading companies toward real coordinated activities.

Governments' and the public's correct understanding of the situation is of course a prerequisite. However, there are difficult issues concerning the government's policy, theoretically and in political reality. As a conclusion, it seems that governments and the public should change some of their current sponsorships for semiconductor manufacturing technology R&D from just a national one to a global one, as far as peoples and industries throughout the world enjoy the benefits of semiconductor evolution.

Conclusions and some implications: the case of semiconductors and beyond

In the semiconductor industry, a need for R&D Test and Evaluation resources and for sharing uncertainty and accessibility about various manufacturing technologies is growing. This puts pressure on companies to join some of their own resources and activities, as seen in the Japanese case. Today such needs make the most decisive reason for the creation of national consortia. There are national consortia in many countries and regions like Japan, the US, and the EU. Economically speaking, however, today's situation as described previously is making the nationality of 'national' consortia more and more irrational. From the perspective of public policy, it seems proper to change the appropriate parts of each current project for semiconductor manufacturing technology R&D from only a national one to a global one.

Concerning the Japanese government's S&T policy, it appears that the government should reconsider and change its R&D project policy to the above end, especially in MITI's Super Advanced Electronics Technology R&D Promotion project. This project should become an example of the integration of the objectives of national competitiveness and international contribution, and the government should gradually increase such cases in its R&D project policy under the expected budget constraints of the twenty-first century.

NOTES

1 The author thanks the German Institute for Japanese Studies, and Hiroyuki Yasunaga (MITI) (as well as other officials); Hideo Setoya (ASET); Hiroyoshi Komiya (SELETE); Tarō Okabe (SIRIJ & STARC); Akira Nagami (EIAJ); Sanetada Mizumachi (SEAJ); Kazuo Higuchi (JEIDA); Kunio Ono (formerly UCOM); Toshihisa Tomie (ETL-AIST-MITI); Hirō Toyoda (NTT Electronics Tech.); Toshiaki Ikoma (TI Japan); Jun'ichi Shimada (Real World Computing Technology Association); Shoji Tanaka (International Superconductivity Technology Center); Dennosuke Uchida (SEMI-Japan); as well as many other people from NEC, Toshiba, Hitachi, Fujitsu, Mitsubishi, Oki, Sony, Matsushita, Sharp, EIAJ, etc. I would also like to thank American colleagues from SIA, SEMATECH, SRC, SEMI, DARPA (DoD), NIST (DoC), Sandia and Los Alamos National Labs. (DoE), NASA, Institute for Defense Analyses, RAND, Intel, Texas Instruments, Bell Lab., IBM, UC-Berkeley (BRIE), Stanford Univ., Univ. of New Mexico, etc. Since this paper is mainly concerned with the Japanese situation, their names are omitted here.

2 However, it should be noted that it is hardly possible to prove this trend, even if one could know the correct amount of R&D expenditure in the industry for a semiconductor generation.

3 It may be difficult to estimate how valuable or how costly for an equipment supplier the input of technological information and needs-specification from a semiconductor manufacturer is. There may be some cases the value a semiconductor manufacturer presented is larger than the earnings from its technical assistance to equipment suppliers. The market mechanism does not work well here.

4 I300I stands for International 300mm Initiative, an international joint venture that is dominated by SEMATECH.

5 For a detailed discussion of the Law and the Plan, refer to Chapters 1, 2, and 5 of this volume.

6 J300 stands for Japanese Virtual Association for 300mm Wafer and related equipment 'standardization' among EIAJ, SIRIJ, JEIDA, SEAJ, and JSNM.

REFERENCES

Inoue, Koki (1996) *Presentation at Kikai shinkō kyōkai-Economic Research Institute's 'Tōkyō Seminar' on 21 May 1996*. (Based on author's research about 10 Japanese semiconductor manufacturers.)

Nihon Denshi Kogyo Shinko Kyokai (JEIDA) (1995) *Denshi kōgyō kisō gijutsu no kenkyū kaihatsu taisei ni kansuru chōsa kenkyū hōkokusho III: Jinzai ikusei oyobi kenkyū shien no taisei ni tsuite* (Report about Japanese R&D posture for the electronics industrial basic technologies III: Nurture of human resource and engineering support). Tokyo: Nihon Denshi Kogyo Shinko Kyokai.

Semiconductor Industry Association (SIA) (1994) *The National Technology Roadmap for Semiconductors.* San Jose: SIA.

WSTS (World semiconductor trade statistics council) (1996 autumn) *World Semiconductor Trade Statistics*. Washington, DC: WSTS.

FURTHER READING

Handotai Sangyo Kenkyusho (SIRIJ) (1995) *Handōtai sangyō: Sōzōteki kenkyū kaihatsu e no chōsen, handōtai sangyō kyōdō kenkyū kikin no teian* (Semiconductor industry: challenge for creative R&D, a proposal for a common semiconductor research fund). Tokyo: Nihon Denshi Kikai Kogyokai.

—— (1996) *21 seiki ni muketa handōtai gijutsu kiban no kōchiku, Handōtai sentan technolojīzu setsuritsu no teian* (Construction of the technology infrastructure for semiconductors toward the twenty-first century, a proposal for semiconductor leading-edge technologies creation). Tokyo: Nihon Denshi Kikai Kogyokai.

Handotai Seizo Sochi Kenkyukai (1995) *Handōtai seizō sōchi sangyō no genjō to shōrai tenbō* (Status and the perspective of the semiconductor manufacturing equipment industry [of Japan]). Tokyo: unpublished research paper by the voluntary member companies of SEAJ.

Kikai Shinko Kyokai Keizai Kenkyusho (KSK Economic Research Institute) (author Koki Inoue) (1996) *Minkan R&D renkei no tenkai to seifu R&D seisaku no yukue, Handōtai kanren sangyō ni okeru kokunai/nichibei mondai* (Status and issues of corporate R&D alliances and governmental R&D project policies, domestic and US-Japan international issues in semiconductor-related industries). Tokyo: Kikai Shinko Kyokai Keizai Kenkyusho.

Nihon Denshi Kikai Kogyokai (EIAJ), Handotai Chuki Bijon Iinkai (1994) *Handōtai sangyō no genjō to shōrai tenbō chōsa hōkokusho* (Report of the status and perspective of the semiconductor industry [of Japan]). Tokyo: Nihon Denshi Kikai Kogyokai.

—— (actually prepared by SIRIJ) (1995) *Heisei 6 nendo handōtai sangyō ni kan suru chōsa kenkyū hōkokusho* (Report on Japanese semiconductor industry in 1994: Toward the future development). Tokyo: Nihon Kikai Kogyo Rengokai.

—— (actually prepared by SIRIJ) (1996) *Handōtai sangyō no shōrai tenbō to chōsa jigyō hōkokusho: Virtual-na dennō shakai ni okeru shin sangyō kōzō* (Report about the future vision of semiconductor industry: Toward a new industrial structure in the virtual electronic brain society). Tokyo: Nihon Kikai Kogyo Rengokai.

Nihon Denshi Kogyo Shinko Kyokai (JEIDA) (author Koki Inoue) (1996) *Beikoku-seifu Dual-Use gijutsu kaihatsu project to kokuritsu kenkyūsho no yakuwari, Handōtaio chūshin to shite* (US government's R&D project policies for dual use technology and its national laboratories under re-consideration – with a focus on semiconductors). Tokyo: Kikai Shinko Kyokai Keizai Kenkyusho.

Sangyo Kozo Shingikai, Sogo Bukai – Sangyo Gijutsu Sho-iinkai (Industrial structure advisory council, general department – industrial technology subcommittee), and Sangyō gijutsu shingikai, sogo bukai – kikaku i'inkai (Industrial technology advisory council, general dept. – planning committee) (June 1995) *Gōdō kaigi hōkokusho* (Joint report of the MITI's industrial structure advisory council's industrial technology subcommittee and the MITI-AIST's industrial technology advisory council). Tokyo.

Science and Technology Basic Law (Law no. 130 of 1995. Effective on 15 November 1995) and *Science and Technology Basic Plan* (Cabinet Decision 2 July 1996). (Unofficial English translation of both can be read at http//:www.sta.gov.jp /.)

12

JAPANESE NANOTECHNOLOGY

Tanya Sienko

Introduction

Nanotechnology has great ramifications for all areas of manufacturing and production and is an extremely interdisciplinary field. 'Strong nanotechnology' refers to the parallel yet individual control of individual atoms and molecules through such devices as self-replicating machines or nanobots. 'Weak nanotechnology' refers to anything falling short of this goal and includes work in constructing nanostructures through semiconductor processing, biotechnology, uses of scanning tunneling microscopes (STM) and atomic force microscopes (AFM) for manipulating individual atoms, and computer simulation of any of the above. In addition, 'nanotechnology' has also been used to refer to work done on a micron scale which should be classified into the realm of micromachines.

Japanese nanotechnology, as it is now proceeding, is almost completely an outgrowth of work in semiconductor processing (nanostructures), micromachines, and biotechnology. It has been pointed out that the development of nanotechnology has been proceeding along two paths: either by (1) taking a piece of material and whittling it down to the required dimensions (so-called 'top-down' construction), or (2) building up from the bottom, with each atom individually positioned, usually through self-assembly techniques ('bottom-up'). Micromachine technology still deals with the former, and it is difficult to see how it will cross over to the latter, which is in the realm of chemical engineering and many of the biosciences. On the other hand, much of the development of mechanical devices/instrumentation (table-top factories,[1] SEMs,[2] waldos,[3] etc.) will be applicable toward bottom-up nanotechnology work as well.

This paper attempts to set out an overview of what sort of nanotechnology research is being accomplished by which organizations in Japan, the ties between those organizations, and finally hint at future developments as they are reflected in the predictions from the Fifth Forecast Survey[4] on when certain nanotechnology 'milestones' will be reached.

Science policy organizations and research plans

The Council for Science and Technology (CST, *Kagaku gijutsu kaigi*) provides general advice to the Prime Minister's Office on basic science policy, establishes long-term research goals, and formulates guidelines for the research supported by the various agencies and ministries. CST is considered to be by far the most influential advisory body on science and technology policy. In 1983, CST created a permanent Committee on Policy Matters consisting of fourteen experts who monitor and evaluate new trends in technology and the natural sciences. Every year, prior to finalization of the national budget, this Committee identifies important issues related to science and technology and prepares a set of 'Guidelines for Science and Technology Promotion.' These guidelines are considered by ministries and agencies when they come to draw up their budgets. The Committee on Policy Matters has sub-committees on policy studies, research evaluation, and research projects. The latter subcommittee supervises the Special Coordination Fund for Promotion of Science and Technology. CST uses this fund to make substantial adjustments to the research funding patterns of individual ministries and agencies. In addition, the CST sometimes develops its own policies in specific areas of science and technology which involve a number of different ministries and agencies (for example: Human Frontiers Science Program).

The Prime Minister's Office receives advice from the Science Council of Japan (SCJ, *Nihon gakujutsu kaigi*), as well as science policy advice on specific research areas from eight other bodies which plan overall policy and coordinate the activities of government institutions concerned with their respective fields.

Although the Science and Technology Agency supposedly has responsibility for ensuring that coordination and balance are achieved in all research not sponsored by the Ministry of Education, Science, and Culture (MESC), in practice it has difficulties in doing so due to the more powerful ministries competing for control of particular areas of science and technology. In any case, each agency and ministry is responsible for its own science and technology program, and receives advice from various bodies associated with each organization.[5]

In spite of the aforementioned policy structure which would seem to indicate a micromanaging of research policy, concrete proposals for individual research projects actually are generated in a much more 'bottom-up' fashion. Usually one or a few well-known researchers have the initial idea and take it to the parent organization. This is then 'investigated' unofficially through 'study groups' (*benkyō-kai*) which incorporate people from industry, government, and academia and act both as opportunities to brainstorm and to provide criticism. At a certain point the plan crystallizes, is written down, and starts on the route of trickling up and down through the bureaucracy to be accepted or rejected.

Japan's 'plans' for nanotechnology

It cannot be said that Japan has a generalized plan or policy for the development of nanotechnology. It is in fact questionable that the full implications of what would be possible even at the more advanced levels of weak nanotechnology have been realized. On the other hand, various projects at different levels are being carried out by different governmental and private organizations, with the implications of the goals of the individual projects being understood very well indeed. The term 'nanotechnology' as a field covering manipulation of material on the nano-length scale in both biological and inorganic fields is used only in Europe and the US. In Japan, 'nanotechnology' is taken to refer at present to the construction of nanostructures on semiconductors and other inorganic surfaces only, although attention is starting to be paid to the biological side. At present, the semiconductor/inorganic efforts are driven mainly by the consortia (government and business) investigating future technology for computers. There has not yet been, however, the fusing together of biotechnology, computer simulation, organic chemistry, and STM/AFM applications which is so characteristic at present in the United States of 'wet' (i.e. bio/organic) nanotechnology research.

At present, 'nanotechnology' is not considered to be an 'important' technology which is under consideration by any of the major policy bodies. Nor is it considered an area whose progress should be tracked, although this may be changing.[6] As mentioned, 'nanotechnology' is considered equivalent with nanostructures (usually on semiconductors and other inorganic materials), with the tracking of research in this area being quite adequately carried out by companies.

For nanotechnology research in Japan the public sector players are three: the Ministry of International Trade and Industry (MITI), the Ministry of Education, Science, and Culture (MESC), and the Science and Technology Agency (STA). Other ministries like the Ministry of Agriculture, Forestry, and Fishery (MAFF) or the Ministry of Health and Welfare (MHW) are involved primarily due to their activities in biotechnology. Private sector players are mainly the large electronic corporations (NEC, Hitachi, *et al.*) together with other chemical- and pharmaceutical-related companies from the biotechnology side.

MITI projects

The first set of nanotechnology-related projects – and perhaps the most applied ones – is being carried out by the Ministry of International Trade and Industry under its Industrial Science and Technology Frontier Program (ISTF). These are projects dealing with research in either obviously important areas (for example non-linear photonics materials), or leading-edge technology. Each project lasts for three to twelve years, and receives anywhere from \$2 million to \$200 million of funding. For each of the individual projects of the program, there exists a public/governmental partner and a private partner, usually a consortium set up by private industry.

The idea has been to promote (1) leading edge research and (2) the cooperation of researchers from academia, industry, and government laboratories. Different projects have had relative levels of success along these lines – Joint Research Center for Atom Technology (JRCAT, see below) has been the most successful at incorporating researchers from all three sectors working together in the same laboratories, while other projects have either had researchers remain at their individual laboratories while sharing the research outcome, or a relatively limited form of personnel exchange. During interviews with managers of ISTF projects, the bureaucratic difficulties involved in incorporating researchers from academia were often mentioned. At present, there is a feeling that although the system is starting to change, it will take several years before researchers from academia will be involved to the same extent as researchers from the other two sectors.

The background of these projects formed a system of budget sponsoring for the 'Research and Development of Basic Technologies for Future Industries' (JISEDAI Program), which was inaugurated in 1981 by MITI and aimed to develop revolutionary technologies essential to the establishment of new industries. The objective fields covered new materials, biotechnology, and new electronic devices, and by April 1988, fourteen themes were under way. Several of the projects were still in the planning stage when MITI decided to incorporate them into the Industrial Science and Technology Frontier Program, started under MITI's newly formed New Energy and Industrial Technology Development Organization (NEDO). The following description of the nanotech-related projects is based on AIST (1995).

The Advanced Chemical Processing Technology project (1990–1996) aims at ultra-high-purity separation and processing technology using a laser beam to selectively excite particles, which then can be pulled down onto a charged substrate; ultra-fine-grained crystal controlling technology which allows the manipulation of particles in a plasma to produce laminated sheet materials or gradient materials; the synthesizing of high-performance organic materials through controlling their formation under ultra-high pressures, strong magnetic fields, and ultra-low temperatures.

By introducing the concept of 'hyper-organized structure controlling,' that is the simultaneous control of structural elements at diverse levels from the atomic-molecular scale to the macro scale, a second project (1991 onwards, first phase five years) attempts to create Synergy Ceramics, a new family of advanced ceramics in which diverse properties are made compatible, and different functions are integrated in the same material. Here the structure of the material is still kept relatively simple. The development of 'intelligent materials' which react to changes in their surroundings is not a goal of this project. The Synergy Ceramics project was sparked by 'futuristic' ideas generated during the Fine Ceramics project carried out under the JISEDAI Program and was pushed very much by industry. Quite a few of the target technologies have obvious, immediate commercial applications.

Under the Molecular Assemblies for a Functional Protein System project (1989–1998), proteins suitable for artificial reconstruction are screened from protein assemblies in biomembranes. The basic conditions necessary to analyze the functions and structures of the proteins will be established and the elementary technologies necessary for artificial reconstruction of functional protein assemblies will be determined. Finally, working protein assemblies will be reconstituted, and their function and structure will be evaluated. So far, the protein involved in a photosystem reaction center utilizing solar energy has been artificially improved and stabilized. A world first was the production of a large amount of this protein using genetic engineering on *Escherichia coli*. Second, the ion channel receptor protein which controls a reaction through molecular recognition has been genetically determined and a technique for producing a large quantity of the protein has been developed.

A fourth project concerned with Quantum Functional Devices (1991–2000) is preparing the next generation computer chip:

> Stepping into the mesoscopic realm in fabrication size, conventional semiconductor devices can not work well due to the appearance of quantum phenomena. In order to overcome the problem, it will be necessary to control quantum effects such as electron tunneling, electron wave interference, and energy level quantization and to pursue operational principles of devices based on quantum effects. The purpose of this project is to establish basic technologies for developing innovative devices with ultra-high speed and multi-functions by utilizing quantum effects appearing in ultraminute structures.
>
> (Rehn 1995: 1)

As for results so far, they have fabricated TiOx quantum wires of 18 nm width using a scanning tunneling microscope, as well as proposed and ascertained the basic operation of several kinds of quantum transistors. The public sector partner for this project is the Electrotechnical Laboratory (ETL) at Tsukuba, while the private sector partner is the Research and Development Association for Future Electron Devices. This association has earlier carried out the just-completed Bioelectronic Devices Project (1986–1995). Several of the topics covered by this research (Molecular Assembly-Based Neural Devices) were related to nanotechnology, although future goals were more along the lines of developing molecular computers.

Probably the most 'nanotechnologized' of all of the MITI projects is aiming at the Ultimate Manipulation of Atoms and Molecules (1992–2001). The purpose of this research is to develop technology for exactly observing and identifying atoms or molecules, and arranging them in a desired layout.

In combination with mechanical probe techniques and beam techniques, the new technology allows the identification, observation, measurement and manipulation of atoms and molecules on the surface of various materials, organic molecules such as DNA, and atomic assembly in free space. The development of simulation technology will also be pursued to exactly predict atomic and molecular processes. During 1994, it was found possible to manipulate structures down to the atomic level by means of magnetic fields. This suggests the possibility of creating new materials through the control of the materials' structures at atomic and molecular levels.

Although this last project appears to be the most 'nanotechnological' project at present, it has been insisted that the main idea is the manipulation of individual atoms with an eye toward creating materials with new properties. The original impetus for the project came from a small band of researchers at the Electrotechnical Laboratory, who approached MITI with the concept which in turn found interested partners in the private sector. Supposedly the ETL researchers' ideas were sparked by Masakazu Aono's Atomcraft project carried out under ERATO (see below). Original plans were for this to be one of the 'large-scale' projects run under MITI, but with the development of the ISTF Program, the decision was made to incorporate it as one of the ISTF projects. MITI seems to be very insistent on attempting to bring together national laboratories, academia, and the private sector. This project has the National Institute for Advanced Interdisciplinary Research (NAIR) as the public sector partner, and the Angstrom Technology Partnership as the private sector partner. A list of the individual members of this Partnership (private partners) shows all of the larger Japanese semiconductor companies, as well as a few from the US and Korea. Perhaps the most interesting partner is Molecular Simulations, Inc. (formerly Biosym), which is a US company known for its simulation software, particularly for biotechnology and pharmaceutical development.

Both NAIR and the Angstrom Technology Partnership have come together to form the (JRCAT) to carry out the above-mentioned research. Although universities are not officially associated with the project, several of the researchers are graduate students (mainly from Tsukuba University) or professors working part-time at national laboratories. The research groups involved are identical with those in NAIR (Table 12.1).

None of these programs should be considered to be on the level of the Fifth Generation computer project. MITI at present seems to have two areas of great interest: micromachines and future computer chip technology. In both cases large scale projects are being carried out such as MITI's ISTF project 'R&D of Micromachine Technology' and various consortia for researching future leading edge computer technology.[7] Both of these are in areas of which there is an obvious market need and (also) obvious applications. At the same time, looking over lists of what is being done and what is considered to become possible with micromachines – such as manipulation

Table 12.1 Groups and targets of the ultimate manipulation of atoms and molecules project of MITI

Group	
Tokumoto	Measurement and control of atomic level structures by mechanical probe
Ichikawa	Observation and formation of atomic scale structures using beam technology
Ozeki	Measurement and control of surface reactions for nanostructure fabrication
Yao	Atomic level analysis and control of II–VI semiconductor surfaces
Tokura	Exploration of transition metal oxides and organic molecular systems
Kanayama	Formation and control of clusters in ion trap and on solid surfaces
Okada	SPM and optical analysis for DNA and organic molecular structures
Terakure	Organic molecules and solids; new techniques for computer simulation
Uda	Semiconductor materials
Hamada	Transition metal compounds
Targets to achieve by the final date (2001) include (intermediate goals in parentheses):	
(Control of local surface reactions)	Manipulation of atoms/molecules
(Control of subnanometer structures)	Control of bulk properties
(Observation and control of growing surfaces)	Formation of superstructures
(Control of reactions in atom clusters)	Formation of nanometer structures
(Observation of molecules)	Molecular fabrication
(Simulation based on first principle calculations)	Reaction system simulation

Source: Author's composition.

of DNA – one cannot help but feel that a lot of the technological predictions that in the US and Europe are associated with the development of nanotechnology are in Japan linked with the development of micromachines. In fact, the US is starting to see the uses of micromachines as 'intermediate steps' and auxiliary equipment for nanotechnology development. In this context, mention should be made of the Nanotechnology Conference held in San Diego, 8–9 December 1996, where one talk linked quite explicitly the use of micromachines and nanotechnology development.

Science and Technology Agency projects

The Science and Technology Agency sponsors two organizations which undertake research into nanotechnology-related fields. Most of the nanotechnology-related projects are under the ERATO Program. ERATO was formed in

1981 in an attempt to stimulate the Japanese research establishment and encourage more bright young researchers into leading-edge research at the beginning of their careers (JRDC 1995; for a detailed discussion of ERATO refer to Chapter 6 of this volume).

ERATO projects are five-year projects, each with a total budget of 1.5–2.0 billion Yen. Although labeled as 'interdisciplinary,' each project focuses on one research topic, under one well-known research scientist. ERATO is more a mechanism for top-class, young researchers to concentrate on particular research areas than an institute by itself, since research is carried out in rented facilities in universities, government labs, and industry. So far, nanotechnology-related quantum device research projects have been the following:

- Quantum Wave Project (1988–1993) led by H. Sakaki (now at IIE, RCAST). Quantum wires, island-type quantum structures, turnstile devices.
- Atomcraft Project (1989–1994) led by M. Aono of RIKEN (now working with NEC groups). Single atom manipulation with STM, room-temperature Coulomb blockade.
- Electron Wavefront Project (1989–1994) led by A. Tonomura of Hitachi Basic Research Laboratory. Electron holography, movement of fluxons in superconducting films.
- Quantum Fluctuation Project (1993–1998) led by Y. Yamamoto of Stanford University and NTT. Uncertainty principle, quantum non-destructive measurements, single electron control.
- International Joint Project on Atom-Arrangement: Design and Control for New Materials (1989–1994) supervised jointly by B.A. Joyce and D.D. Vvdensky of Imperial College (UK) and T. Kawamura of Yamanashi University.

The other organization supported by STA is the Institute of Physical and Chemical Research (RIKEN). Known for its international flair and interdisciplinary work, three of its laboratories out of a total of twenty are especially noted for their nanotechnology slant: the Laboratory for Nano-Electronics Materials, Laboratory for Nano-Photonics Materials, and Laboratory for 'Exotic' Nano-Materials. They are part of the Frontier Materials Research Program. So far, their research seems to have been geared toward construction and characterization of quantum device structures (Table 12.2).

Projects under the Ministry of Education, Science, and Culture

MESC provides money to Japanese universities, so any direct university research in nanotechnology falls under its jurisdiction. At present, the major

Table 12.2 Nanotechnology-related laboratories of RIKEN

Laboratory for Nano-Electronics Materials T. Sugano, Head Characterization and control of surface and interface processes Nanoprocessing Electrical and optical properties of nano-electronics materials Development of nano-electronics materials and nanostructures research to device applications
Laboratory for Nanophotonics Materials H. Sasabe, Head Creation of QW structures of low-dimensional conjugated compounds Elucidation of dynamic-behaviors in excited states by ultrafast spectroscopy Study on photorefractive index change in photonics materials 2D crystallization of photoresponsive proteins and physical properties Creation of optical neural networks with self-feedback function
Laboratory for Exotic Nanomaterials W. Knoll, Head Studies of nanoscopic fabrication, characterization, modification Studies of high resolution electron microscopy
Related work at RIKEN includes: *Plant homeostasis research* Laboratories for photo perception and signal transduction, for plant hormone function, and for plant biological regulation *Glycobiology research* Laboratories for glyco-cell biology, for molecular glycobiology, and for glyco-technology *Research on brain mechanisms of mind and behavior* Laboratories for neural information processing, for synaptic function (neural networks), and for neural systems *Research on brain information processing* Laboratories for neural modeling, for information representation, and for artificial brain systems *Photodynamics research* Laboratories for submillimeter waves, for photophysics, for organometallic photodynamics, and for photobiology *Bio-mimetic control research* Laboratories for neural circuits, for genes of neural systems, for bio-mimetic sensory systems and for bio-mimetic control systems

Source: RIKEN (1996).

interdisciplinary research center under MESC is the Research Center for Advanced Science and Technology (RCAST). This center, part of Tokyo University and located at its Komaba campus, is one of the offshoots formed upon the reorganization of Tokyo University's Institute of Space and Aeronautical Science in 1981. Among the twenty-two areas of research are five related to nanotechnology (Table 12.3).

Many of the projects listed in Table 12.3 have been carried out in conjunc-

Table 12.3 Nanotechnology-related RCAST projects

Advanced devices dept.
Quantum microstructure devices (Prof. H. Sakaki – also connected with ERATO project)
Molecular beam epitaxy and ultrafine lithography for quantum wells, wires, and boxes
Scanning tunnelling microscopy of quantum microstructures
Transport study of quantum microstructures and ultrafast electronic devices
Optoelectronic and spectroscopic studies of quantum microstructures, lasers and other photonic devices

Advanced materials dept.
Photonic materials (Prof. Y. Shiraki)
Atomically controlled growth technology of semiconducting materials for photonic devices
Fabrication and characterization of photonic devices
Physics of mesoscopic and low-dimensional electron system

Advanced systems dept.
Nanometer-scale manufacturing science (Prof. T. Suga)
Ultra-precision machining of advanced materials
Surface-activated bonding of dissimilar materials
Interconnections and packaging of microelectronic components

Biomedical devices dept.
Future goals include 'development of micromachines' and 'nanomachines for biomedical use' (Prof. I. Fujimasa)

Life engineering dept.
(being established)

Source: Author's composition.

tion with groups at Tokyo University's Institute of Industrial Engineering. Also not to be omitted is Prof. Karube's work in bioelectronics, which deals with 'organic sensors on a chip.'

Besides the above work, many individual universities are undertaking research projects which can be considered part of nanotechnology (quantum effect devices and mesoscopic structures, biotechnology, advanced and improved instrumentation), although there appears to be no claim to be doing research on 'straight' nanotechnology. The universities which seem the most active in these areas are the Tokyo University's Institute of Industrial Engineering, Tokyo Institute of Technology, Osaka University (bioscience), and Tohoku University. Tokyo Institute of Technology has just established a new department for bioscience; it is also where quite a few quantum effect device research projects are based. Tohoku University, aside from other projects, is starting more STM-based work and is the major player in micromachine-related research. This is by no means a complete list. Collaborative work is also increasing, such as the joint project between NEC researchers and a team at Tokyo University, who have found a way to apply computer-generated holograms to the fabrication of nanostructures (Electronic Engineering Times 1996).

Private sector efforts

As already mentioned, many of the large electronics firms are involved in the Angstrom Technology Partnership as well as in the Research and Development Association for Biotechnology and Chemical Industries are equivalently associated with the Research Association for Biotechnology – connected with the Molecular Assemblies for a Functional Protein System project of MITI. Individual research efforts, too many to list, are also being carried out. In passing it should be mentioned that the two 'quasi-companies' NTT and ATL (Advanced Telecommunications Laboratories) are carrying out very interesting research, albeit in different areas. In nanotechnology-related work, NTT's Atsugi laboratories are working on Coulomb effect devices, using STMs to create holes in silicon, and self-organization of growth on strained materials to produce quantum dots. ATL is more on the 'information' side of research, working on evolutionary programming, cellular automata, and virtual reality – Tom Ray and his computer program Tierra demonstrating evolutionary life; Hugo de Garis and his 'growing an artificial brain.' Another researcher, Hitoshi Hemmi, has been working on evolutionary programming techniques to 'evolve' hardware circuits (ATR 1994, Mizoguchi, Hemmi, and Shimohara 1994).

Hitachi's Advanced Research Laboratory has carried out the construction of an experimental apparatus that many researchers would consider to be the ultimate scanning tunneling microscope; namely, building a Transmission Electron Microscope into the device so that observations of the STM needle in situ can be carried out to extremely high resolution. The STM in question is actually a 'STM on a chip' formed by standard dry etching of a silicon wafer, and is the brainchild of Mark Lutwyche and Yasuo Wada (Lutwyche and Wada 1995). Already, Wada's research group has been able to demonstrate conclusively that an atomically sharp STM apex results in STM images with atomic resolution. They have also demonstrated how to grow very sharp nano-needles on the substrate surface. The next attempt, according to Wada, will be to fuse his Future Electron Devices. Furthermore, it seems reasonable to assume that Japanese groups work with that of the electron tomography group such that in situ electron tomography is also possible, thus giving both surface- and inner-measurement facilities while the STM is operating.

Another of Wada's ideas, more futuristic, is of atom/molecular switching devices, Atom Relay Transistors (ART) and Molecular Single Electron Switching (MOSES) devices. These devices would have total dimensions well below a few nanometer and an operation speed of more than a terahertz (10^{12} Hz). The basic ART configuration consists of an atomic wire, a switching atom, a switching gate, and a reset gate. When the switching atom is displaced from the atom wire by an electric field (supplied by the switching gate), the ART turns off. Memory cell and logic gates have also been conceived. A supercomputer based on such devices with 10^7 logic gates and 10^9 bits of memory would fit in an area of 200 square microns and operate at terahertz switching speeds.

Wada's proposed molecular devices (MOSES) are actually a form of Single Electron Tunnelling devices. Here, metal and semiconductor are replaced with conducting and insulating polymers, respectively. Simulations have been carried out assuming polyacetylene and polyethylene. Polydiacetylene (a polyringed molecule) is another possible candidate if one of triple bonds linking its rings is changed to a single bond to produce a tunnelling gap (Wada 1995).

Conclusion: future prospects and final hurdles

Japanese scientists do see a bright future for their work. One of the projects that the National Institute of Science and Technology Policy (NISTEP) is involved in is a full-scale technology forecasting survey using the Delphi technique, repeated every five years. Roughly 1,000 topics divided among fourteen fields are sent out to over 3,000 researchers considered to be experts in individual fields. Estimated time of realization, importance, necessity of international cooperation in research and development, and so forth are requested on each topic. After collating all answers, the questions are sent out again, along with the estimates from the first round. Researchers are free to modify their answers based on knowledge of other people's expectations. The nanotechnology related issues are summarized in Table 12.4.[8]

The roadblocks which exist in the path of nanotechnology development in Japan are far more due to the structures of how technology is developed, rather than any actual technical barriers. Japanese research laboratories are still structured along a very vertical system (*kenkyūshitsu*), which leads to isolation and lack of contact between groups even in the same building. Also, so far Japan has shown little mixing between fields. This may be changing, with the emergence of laboratories like NAIR, but for the most part researchers working in semiconductor processing (for instance) have been working completely independently of researchers working in chemistry. The main split that still seems to divide Japanese researchers from their Western brethren is the relative lack of contact between the organic sciences and inorganic sciences. Since some of the most interesting advances in nanotechnology have been occurring in the organic chemistry/biotechnology sciences, this is a potentially fatal gap.

Another potential difficulty occurs, if pursuing nanotechnology research from the biotechnology side seems to be more productive, because this leads into the relatively low level of advancement (by comparison with the US and

Table 12.4 Nanotechnology-related results of Delphi Surveys

Mass production of new materials constructed by using ion and particle beams with controlled characteristics (J-2006, G-2010)

Elucidation of adhesion mechanisms of metal-polymer interfaces (J-2002, G-2002)

Use of processing technology for super-smooth metal mirrors (J-2003, G-2002)

Embedding impurities and repairing crystallized silicon surfaces with STM technology (J-2003, G-2004)

Organic hybrid composite materials of controlled structure at the mono-molecular layer level (J-2003, G-2005)

Widespread use of Atomic Layer Etching in semiconductors (J-2003, G-2004)

Organic and inorganic composite materials with constituents in size of several to several ten nanometers (J-2004, G-2004)

Development of polymer processing technology for controlling microstructures in 1 to 10 nm size (J-2005, G-2007)

Possibility to control the structure and properties of solid interfaces at the atomic level (J-2005, G-2006)

Development of techniques for synthesizing substances with new functions based on the combinations of numerous types of bonds at the atomic level (J-2008, G-2004)

Another set of topics which could be considered nanotechnology-related but which were only investigated in Japan are the following (plus year of predicted realization):

Development of a material having a flexible function equivalent to that of human muscles (2013)

Widespread use of atomic-layer epitaxy and atomic-layer etching in all compound semiconductors (2003)

Development of structural materials which can indicate residual service life (2011)

Development of intelligent materials which incorporate sensor functions, storage functions, and effector functions (2009)

Development of technology enabling complete control of two- and three-dimensional protein structures (2008–9)

Development of integrated circuits achieving advanced functions through self-organization adapted to the given tasks (2011)

Development of technology enabling molecular-level write and read memory of great quantities of information (capacity of at least 10^{12} bits/cm^{2}) (2015)

Sources: NISTEP (1992), (1994).
Note: Japan = J. German = G.

Europe) of Japanese biotechnology. Biotechnology is considered by MITI and other ministries to be one of the potential new technologies for the next century which can provide the production of highly value-added products necessary for Japan's economy. At the moment, biotechnology is overseen (or 'guided') by five separate government bodies (MITI, MESC, STA, MHW,

MAFF). Attempts have been made to consolidate biotechnology under one organization's control but, as of now, they have not succeeded (Salusbury and Otsuka 1994).

A third, and to my mind very threatening difficulty is Japan's lack of experience in computer simulation work, and their lag in computer networks. Particularly for future developments in bioscience and future 'nanochemistry' work, graphic visualization of data and realistic simulation will be the driving edge for future developments in the field. We have already seen in a simple demonstration, the use of a data glove coupled to a screen and an STM in a 'virtual reality' demonstration of picking up a single atom and moving it (Byte 1996: 3). This sort of graphic manipulation will be more and more important in the future. The compilation of material into databases which can be readily accessed by researchers who need it is also a key point, particularly in biotechnology.

In any case, it is doubtful that commercialization of Japanese nanotechnology, when it comes, will occur along equivalent lines as to what will take place in the US. Japanese nanotechnology commercialization on the side of quantum effect devices will undoubtedly be dominated by the giants such as NEC, Hitachi, Fujitsu, Toshiba, and very likely the pharmaceutical companies on the biotechnology side. Japan does not offer much in the way of venture capital or other opportunities for small start-up companies, and the small and medium enterprises are likely to be left behind, at least in the beginning. The advantages of higher density and lower energy-use devices in the electronics hardware industry are obvious enough (and a natural enough extension of present technology) that little resistance is likely to be encountered from that direction. Difficulties occur only when a new technology appears 'spontaneously' which poses a direct threat to the technology presently in use in which the major players have a vested interest. At such a point, if the major players are 'frozen' enough by their inertia, it is possible for a smaller, more nimble company to 'pick up the new technology and run with it.'

In summary, Japan remains 'uncommitted' to nanotechnology in the strong sense of the term (i.e. nanobots, general-purpose assemblers) but Japan does not want to fall behind the rest of the world in the underlying technologies, all of which have ready commercial applications. As mentioned, it is questionable that Japan realizes the ramifications of 'strong nanotechnology.' The question is what will happen when enough advances have accumulated that nanotechnology starts looking like it is actually possible in a predicable time frame. The Japanese system is centralized enough that a full-scale, nation-wide, coordinated development thrust could easily be implemented at such a point – and it would not be too surprising if they were to do so.

NOTES

1 'Table-top factories' refer to factories small enough to fit on top of a table.
2 Scanning Electron Microscopes.
3 'Waldos' are robotic arms, linked physically or electronically to gloves, such that the movements of the person wearing the gloves are reproduced either on a much larger scale or much smaller scale by the robotic arms. Term taken from a story by Robert Heinlein.
4 A technology forecasting survey using the Delphi technique.
5 MESC is advised by the Science Council, the University Council, and the Geodesy Council. MITI receives advice on policy and programs from the Industrial Technology Council. The Ministry of Posts and Telecommunications (MPT) has the Telecommunications Council, etc.
6 The author recently gave a well-received talk to a group at the Japanese Bioindustry Association explaining the concepts behind US 'wet' nanotechnology research and their implications. This may lead to a *benkyō-kai* inside MITI in the future.
7 See reports from the Micromachine Center (http://www.iijnet.or.jp/MMC/).
8 These topics were used in a mini-survey investigating technology forecasting in Germany as well – the prefixes indicate whether the year indicated is the average year forecast by German or Japanese researchers.

REFERENCES

AIST (Agency of Industrial Science and Technology) (1995) Pamphlet on Industrial Science and Technology Frontier Program (ISTF). Tokyo: Japan Industrial Technology Association.

ATR (Advanced Telecommunications Research Institute International, Human Information Processing Research Laboratories) (1994) *Research Reports*.

Byte (Peterborough), June 1996, http://www.byte.com/art/9606/sec8/art3.htm.

Electronic Engineering Times, 20 May 1996, Issue 902, http://techweb.cmp.com/techweb/eet/news/news96.html.

JRDC (Research and Development Corporation of Japan) (1995) Pamphlet on ERATO, Kawaguchi-shi, Japan.

Lutwyche, Mark I. and Yasuo Wada (1995) *Sensors and Actuators*, A 48, pp. 127–36.

Mizoguchi, Jun'ichi, Hitoshi Hemmi, and Katsunori Shimohara (1994) *Production Genetic Algorithms for Automated Hardware Design Through an Evolutionary Process*, ATR internal report.

NISTEP (National Institute for Science and Technology Policy) (1992) *The Fifth Technology Forecast Survey – Future Technology in Japan.* Tokyo: NISTEP (= NISTEP Report no. 25).

—— (1994) *Outlook for Japanese and German Future Technology – Comparing Japanese and German Technology Forecast Surveys.* Tokyo: NISTEP (= NISTEP Report no. 33).

Rehn, Victor (1995) *A Brief on the R&D Association for Future Electron Devices*. Office of Naval Research, Asian Office, Tokyo.

RIKEN (1996) Explanatory documents at http://www.riken.go.jp.

Salusbury, Tom and Yasuko Otsuka (1994) Support for Innovation in the Japanese Bioindustries. (Unpublished) *Working Report of the Science & Technology Section*. British Embassy, Tokyo.

Wada, Yasuo (1995) *Optoelectronics – Devices and Technologies*, vol. 10, no. 2, pp. 205–20.

13

JAPANESE R&D ACTIVITIES IN ADVANCED MATERIALS

The case of superconductivity research

Michael Becht

Introduction

The field of advanced materials is generally regarded as being crucial for future technological development. Relatively little is yet known about actual research programs being conducted in the major industrial countries that are related to this field.

This paper addresses Japanese R&D activities in new materials with a particular focus on the area of superconductivity research. In the second section, some introductory remarks concerning advanced materials and their applications are given. Thereafter, the third section contains a brief sketch of Japanese research projects in this field and a comparison of actual and future market size in Japan and the United States; the fourth section provides an overview of Japanese research programs in the particular field of ceramic high-temperature superconducting materials, and in the fifth section the structure, activities, and performance of a major Japanese research institution in this field, the International Superconductivity Research Laboratory (ISTEC), are analyzed. The paper closes with some concluding remarks.

Advanced materials: classification and applications

Materials collectively known as 'advanced or new materials' have begun to attract attention as the third generation of new industrial materials, following in the wake of conventional metals and plastics. They are defined as follows:

> New materials are high value-added materials exhibiting epoch-making properties and producing a level of social value never before experienced, which are made from metallic, inorganic, organic, or a

> combination of such raw materials and are the product of high-level manufacturing/processing technology, or a commercial application technology.
>
> (JETRO 1990: 1–2)

The advanced and new materials discussed in this chapter refer to those high value-added materials which have achieved an improved level of functions and structural characteristics through compensating for the weaknesses and extracting the superior qualities of conventional materials.

Advanced materials are classified into four categories: new metals and alloys, ceramics, special polymers and advanced composites. Another classification is via their functions, which are separated into mechanical, thermal, electrical, magnetic, optical, biological, separation and chemical function groupings. The relationship between these classifications is given in Table 13.1. As can be seen from the table, metals and alloys as well as composites have less functional characteristics than special polymers and ceramics.

Composites as well as metals and alloys are widely used due to their structural characteristics. The latter ones show great strength, malleability and electrical conductivity. Examples of applications are in aircraft (aluminum alloys), gas turbines (nickel base alloys) and magnetic levitation (niobium tin alloy).

Composites consist of a reinforcing material (carbon, Aramid® or aluminum fibers) and a base material (polymers, metals, ceramics). They are superior to the base materials because it is possible to design improved strength and elasticity. Aircraft and space equipment, gas turbines, leisure and sporting goods are their principal fields of application.

Functional characteristics are more important in the case of special polymers and ceramics. In addition to low weight, good workability and high corrosion resistance, special polymers have excellent performance and special functions, e.g. the separation function is displayed in ion exchange resins (styrene, acryl) and membranes, the optical function of liquid crystals is used in displays.

Table 13.1 Classification of advanced materials

Functional characteristics	*Special polymers*	*Ceramics*	*Metals/Alloys*	*Composites*
Mechanical	x	x	x	x
Thermal	x	x	x	x
Electrical	x	x	x	
Magnetic		x	x	
Optical	x	x		
Biological	x	x		
Separation	x	x		
Chemical	x			

Source: JETRO (1990: 2–3).

Finally, ceramics are manufactured from high purity precursor materials in a certain process, enabling control over the microstructure of the material produced. In light of their physical properties and functions, they are further classified as electromagnetic ceramics, engineering ceramics, or bio-ceramics. Their applications are in electronics (substrates, devices, sensors, etc.), in the machinery industry (hard coatings for cutting tools, gas turbines, etc.) and of minor importance in the fabrication of histocompatible products (artificial teeth). It should be mentioned that even in the 1970s, the application of ceramics as structural components was considered very important, especially in the automotive industry. However, their brittleness and the high cost associated with their fabrication prevented widespread use. Successful exceptions are the catalyst used for reducing exhaust gas pollutants (hydrocarbon, carbon monoxide and nitric oxide) and the exhaust gas oxygen sensor (Narula *et al.* 1996: 984–1003).

Japanese research projects and market size in advanced materials

Advanced materials have become one of the main objects of Japanese government R&D programs since the early 1980s. The Ministry of International Trade and Industry (MITI) established the Fine Ceramics Office in 1982 and has launched several programs related to advanced materials between 1981 and 1988 (JETRO 1990: 4–5):

- basic technologies for advanced materials: eight projects with a duration between seven and ten years; a total budget of $795 million;
- large-scale manufacturing technology: one project from 1986 to 1993; a total budget of $150 million;
- the Moonlight Project (generator technology): two projects with a duration of seven and eight years; a total budget of $410 million.

Additional research activities in advanced materials were started also by other organizations that were related to different government institutions. In 1981 the Research Development Corporation of Japan (JRDC) which is affiliated to the Science and Technology Agency (STA) initiated an innovative research effort called Exploratory Research for Advanced Technology (ERATO), for the purpose of fostering the creation of advanced science and technology. This cluster of projects includes several on advanced materials.

For another instance, the Railway Technical Research Institute (RTRI) which belongs to Japan Railways, JR Central, is promoting basic technology for the Maglev (magnetic levitation) system. One main development aim of RTRI is the enhancement of the reliability and durability of the superconducting magnets.

The development of the total market size of advanced materials in Japan has been compiled by the Japan External Trade Organization (JETRO 1990: 11pp.) on the basis of two studies carried out by MITI groups. According to these studies, the market will increase from sales of $31.7 billion in 1987 to $120.4 billion in the year 2000 (Table 13.2). This corresponds to an increase of a factor of four and an average annual growth rate of 10.8 percent. The sales of new inorganic materials (ceramics and diamond), currently holding a share of 36 percent of the total market size of advanced materials, are increasing by a factor of 5.2 from 1987 to 2000. This forecast of sales growth projects a 50 percent share of the total market size then.

Since new inorganic materials are considered to play an important role in the future, some detailed numbers are given in Table 13.2. These materials comprise the subgroups ceramics and diamond. Furthermore, ceramics are divided into six sectors comprising electromagnetic, machine tool/mechanical, thermal, chemical, optical ones and others.

The market size for electromagnetic ceramics is expected to reach $28.87 billion by the year 2000 with an average annual growth rate of 11.1 percent. The total market share of these materials, currently 70 percent of ceramics production, is expected to decline significantly to just over 50 percent by the year 2000. By contrast, the fast-growing structural materials sector may increase its share, with the markets for machine tool/mechanical materials and for thermal materials rising to about $5 billion each. Similar trends are predicted for chemical/medical, optical and other materials.

Table 13.2 Size of the market for new materials in Japan (billion US$)

New materials category	*1987*	*2000*
Special polymers	8.45	21.70
New metals and alloys	6.45	21.80
Advanced composites	5.35	16.80
New inorganic materials		
Ceramics		
1 Electromagnetics	7.40	28.87
2 Machine tool/mechanical materials	1.00	5.04
3 Thermal materials	0.50	4.94
4 Chemical/medical materials	0.62	4.51
5 Optical materials	0.92	7.15
6 Others	0.20	4.65
Subtotal	10.64	55.16
Diamonds	0.82	4.98
Total	11.46	60.14
Grand total	31.71	120.44

Source: JETRO (1990: 11–13).
Note: Conversion rate: $1 = Yen 100.

In a recently published study concerning the US advanced ceramic industry (Abraham 1995: 68–72), data about the present market as well as a forecast of the future markets were included. In this study, ceramic materials have been classified into structural ceramics, electronic ceramics and ceramic coatings. Structural ceramics comprise heat engine components, cutting tools, wear-resistant parts and bio-ceramics. Electronic ceramics are used for their functional characteristics (e.g. insulators, substrates, capacitors, piezoelectric ceramics, ferrite magnetic and high-temperature superconducting materials). High-performance applications where properties such as wear resistance, hardness and corrosion resistance are important, are the features of ceramic coatings.

In Table 13.3, the data for the US market for ceramics is summarized. The total value of the market for 1994 was estimated to be $4.9 billion. This is projected to increase to $8.5 billion by the year 2000, with a yearly growth rate of 9.8 percent from 1994 to 2000. In terms of market share, electronics constitute 79 percent at present, however, as the highest growth rate is for structural ceramics, the market share of electronic ceramics will slightly decrease to 77 percent by the year 2000.

In order to compare the data sets of the Japanese and US studies, some of the data in the JETRO report has been summarized and classified to be consistent with the US terminology (see Table 13.4). The data in the Japanese study has been altered as follows: thermal and chemical/medical

Table 13.3 US market for advanced ceramics (billion US$)

Categories	*1994 (share)*	*2000 (share)*	*AAGR %*
Structural ceramics (SC)	0.47 (10%)	1.02 (12%)	13.5
Electronic ceramics (EC)	3.86 (79%)	6.57 (77%)	9.3
Ceramic coatings (CC)	0.53 (11%)	0.94 (11%)	10.2
Total	4.86	8.53	9.8

Source: Abraham (1995: 68–72).
Note: AAGR = average annual growth rate.

Table 13.4 Market size of ceramics in Japan (billion US$)

Categories	*1994 (share)*	*2000 (share)*	*AAGR %*
Structural ceramics (SC)	1.12 (12%)	9.45 (22%)	17.9
Electronic ceramics (EC)	7.40 (78%)	28.87 (67%)	11.1
Ceramic coatings (CC)	1.00 (10%)	5.04 (11%)	13.3
Total	9.52	43.36	12.4

Source: Compiled from Table 13.2.
Note: AAGR = average annual growth rate.

materials (no. 3 and 4 in Table 13.2) are combined and renamed structural ceramics (SC), electromagnetic materials (no. 1 in Table 13.2) are renamed as electronic ceramics (EC) and machine tool/mechanical materials (no. 2 in Table 13.2) are renamed as ceramic coatings (CC).

The US as well as the Japanese data show that electronic ceramics have the largest share followed by structural ceramics and ceramic coatings. A large growth of the market is expected with the highest average annual growth rates being in the structural ceramics section followed by the coatings and electronics sections. The data in Table 13.3 and 13.4 show, that even in 1987, the Japanese market for ceramics was more than twice as large as the US market in 1994. The expected market growth is also higher in Japan according to the JETRO data. Considering the current situation of new materials, the Japanese forecasts may be too optimistic but reflect the Japanese opinion that opportunities exist to create new business in the field of advanced materials. On the other hand, new materials have commonly been seen by US companies in the context of their existing business (Lastres 1994: 156–7).

Japanese research activities and market potential in ceramic high-temperature superconducting materials

The most promising class of ceramic materials, high-temperature superconducting materials, was discovered in 1986. The Nobel prize was given to Klaus Alex Müller and J. Georg Bednorz of the IBM Research Center in Switzerland in 1987 for their discovery. High-temperature superconducting materials are composed of mixed oxides containing rare earth oxides and copper oxides. Some materials show superconducting properties, which means no ohmic resistance and therefore no loss of energy, up to temperatures well above the boiling point of liquid nitrogen (77 Kelvin or −198°C). Since liquid nitrogen is available in industrial quantities, even large scale applications can easily be realized. Small cooling devices are also currently in the development stage to enable small and portable applications.

Three classes of products are currently being developed:

1 bulk material (polycrystalline materials and single crystals) for magnets and substrates in electronics;
2 tapes and wires for motors and generators (transport of energy, energy storage, energy conversion); and
3 microelectronics devices (junctions, sensors, antennas).

Japanese industry and government research centers have been involved since the early 1970s in R&D programs for low-temperature superconductors as well as for ceramics (Lastres 1994: 103–5). Therefore, the Japanese response to the discovery of high-temperature superconducting ceramics came rapidly, despite the fact that for public research centers, the budgeting system is considered to be rigidly defined and administered by the Ministry of Finance in a way which does not accommodate new discoveries in science.

- The first reaction came in January 1987 from Japanese firms already engaged in low-temperature superconductor research by creating project teams and at the end of 1987 the establishment of specific R&D departments (Hane 1991: 71–5).
- The Ministry of Education extended an existing program of low-temperature superconductors for further three years in March 1987.
- A New Superconducting Materials Forum was established by the STA in the beginning of 1987 and the launching of its Multi-Core Project followed in September.
- In the initial stage, MITI's response to the new scientific discovery appeared to be somewhat slow (Hane 1991: 88–9). But once a decision to launch a related program was made, activities developed very quickly: in December 1987, the establishment of a joint research center was announced and in January 1988 the International Superconductivity Technology Center (ISTEC) was opened.

After 1987, superconductors became the central priority of government R&D programs for advanced materials as summarized in Table 13.5. A primary distinction between these and the earlier programs in this area is the greater importance conferred on basic research. The first three programs listed in Table 13.5 are almost entirely devoted to basic superconductivity research at the scientific frontier. The remaining five are at the state-of-the-

Table 13.5 Japanese R&D projects in superconductivity after 1986

Project	*Sponsor*	*Period*	*Budget (million $)*
ISTEC	MITI	1988–97	250
Multi-core project	STA	1988–92	150
Oxide superconductors	MESC	1987–90	
1 ERATO project	STA	1986–91	
Supercond. devices	MITI	1988–97	250
Supercond. generator	MITI	1988–95	250
Supercond. magnetic energy storage	MITI	1990–99	250
Supercond. sensor.	MITI	1990–96	57

Source: Lastres (1994: 109–10)

art level in their areas with respective targets involving the leapfrogging of existing technologies.

Subsequently, a brief overview concerning the organization of STA's Multi-Core project and MITI's ISTEC will be presented (ISTEC, chosen as a case study, will be described and discussed in more detail in the next section). The Multi-Core project has a very general goal of promoting basic research in new high-temperature superconductors. STA provides the laboratory facilities which serve as the research centers or 'cores,' and provides funds to enhance existing equipment with the intention of conducting 'cutting-edge' research. These centers are the heart of the Multi-Core project, with each core focusing on a different technical theme. The cores are staffed with researchers from the host facility, with these facilities also made available to researchers from other public and private sector organizations who submit proposals for collaborative research. The expertise available at each core and the extensive laboratory facilities are the most important incentives for collaboration. This can be particularly attractive for small and medium sized companies as well as for research groups from universities.

In ISTEC, unlike the other MITI promoted cooperative projects listed in Table 13.5, there are no specific technical targets. ISTEC is a non-profit organization open for any public institution and private company to participate in, including non-Japanese ones. The financial basis is provided by the members, together with MITI's New Energy and Industrial Technology Development Organization (NEDO). The research facilities of ISTEC are located at the Superconductivity Research Laboratory (SRL) in Tokyo which has also two branches in Morioka (Iwate Prefecture) and Nagoya (Aichi Prefecture). The staff consists of researchers from member companies, employed staff and foreign research fellows financed by STA, NEDO, and ISTEC.

As in other Japanese R&D programs, one important aspect has been the explicit emphasis on the monitoring, collection and diffusion of information concerning superconductivity. Both MITI and STA basic research programs have institutionalized supplementary activities aimed specifically at promoting the gathering and diffusion of information within the community formed by the participants.

Ten years after their discovery, a more realistic view of the opportunities for application of superconducting materials has replaced the enthusiasm of 1987. In the beginning, large R&D activities grew extremely rapidly all over the world due to the expected tremendous capabilities of high-temperature superconducting materials. The Nobel Prize was given to Müller and Bednorz only one year after the discovery, which is very unusual but indicates how highly it was considered. However, these expectations have been gradually reduced in the following years. Many of the characteristics of these ceramics were not well understood or even unknown, which drastically slowed down the development of processes and products. Two examples are: (1) the fabrication process (equipment, testing, etc.) for electronic devices which had to be modified from that used in the semiconductor industry to a more advanced one suitable for ceramic materials, and (2) the fabrication process for wires, which require a flexible substrate due to the brittleness of the high-temperature superconducting materials. The development of standards for products as well as for manufacturing processes was difficult due to the lack of knowledge of many aspects of these materials, e.g. their long term stability. In addition, it should be noted that as yet, there is no theory which can fully explain the physics of superconductivity.

Despite all these problems, the enthusiasm in 1987 was important to initiate the R&D projects and to procure funding for the programs. The investment put into these projects has been large in terms of money and time, which has resulted in notable progress through the years and the first products are now on the market or in the testing stage. The first major market for high-temperature superconductor technology is believed to be in magnetic resonance instruments. These instruments currently comprise well over US$ 1 billion a year market, whose customers are in the healthcare and chemical industries. High-temperature superconducting materials are the enabling constituents of a new receiver probe technology improving both instrument sensitivity and data collection speed at modest incremental cost to system users (Simon 1995: 18–19). Power applications such as fault current limiters and wires for power transmission cables and microwave devices are considered to be the next candidates (Lubkin 1996: 48–9; Kenward 1996: 29–30).

Certainly, the existing number of successful results is smaller than that expected ten years ago. However, the appearance of first products on the market clearly indicates the potential of high-temperature superconductors. There are numerous opportunities for products, but public funding is still required to carry on research as well as to stimulate new projects in this field. As a result, budgets for public R&D programs related to superconductivity are still very high in the US and Japan. In the US, expenditure was around $200 million per year from 1993 to 1995 (MK Associates 1995). The Japanese budgets for 1994 and 1995 were in about the same range (ISTEC 1995: 71–2).

The International Superconductivity Research Laboratory (ISTEC): a case study in Japanese superconductivity research

Outline

ISTEC is one of the Japanese projects in the field of superconductivity guided by MITI. It is a ten year project, which began in January 1988, funded by its members (companies and public institutions) and NEDO. The shift of MITI's policy in R&D programs in the 1980s toward the promotion of basic science is clearly reflected in the rather broad target of ISTEC, i.e. contribution to the progress of superconductivity and development of superconductivity-related industry.

In 1996, the enthusiasm of certain companies to further participate in ISTEC has decreased and also, critics concerning the lack of broad success (see the previous section) makes the present situation difficult for ISTEC. Nevertheless, some excellent results have been presented during the years and about 355 patents (to April 1996) have been filed. Presently it appears that there will be an extension of the program, but some reorganization may also occur.

Major functions

ISTEC's activities are defined in the following fields:

- Conducting surveys and studies of global trends in superconductivity technologies and on the feasibility of their application in various industrial fields.
- Educating and disseminating information by organizing and attending symposia, seminars and study groups; publishing of the ISTEC Journal and other reports on superconductivity.
- Promoting international exchange by the hosting of workshops, accepting foreign researchers and dispatching Japanese researchers to organizations overseas.
- Conducting basic research and development of the theory of high-temperature superconductivity, materials exploration and innovative processing.

The first three functions are performed by ISTEC and the last by its associated Superconductivity Research Laboratory (SRL). This institutionalized separation of functions clearly reflects the policy of MITI to give the researchers space to concentrate on their research work. ISTEC fulfills the above objectives by:

- Publishing reports on the results of surveys and studies on the feasibility of the application of superconductivity technologies in fields such as transportation, electric power, electronics, medicine and accelerators (in both Japanese and English).
- Publishing various technical reports on the results of the research done at SRL (in both Japanese and English).
- Hosting the 'International Symposium on Superconductivity', an annual symposium where researchers of various fields related to superconductivity participate (in English); the contributions are published in symposium proceedings.

- Hosting study meetings on superconductivity for technical and inter-industrial exchanges (in Japanese).
- Hosting the 'International Workshop on Superconductivity' annually; this workshop is related to a specific topic in high-temperature superconducting materials research (in English).
- Cosponsoring the 'Superconductivity Industry Summit' together with the Council on Superconductivity for American Competitiveness (CSAC) and the Consortium of European Companies determined to Use Superconductivity (CONECTUS).

It should be noted that the main language is English which reflects the trend of internationalization in Japanese R&D projects. However, a large number of reports are initially only published in Japanese and the English translations usually have a delay of several months.

Members and budget

ISTEC is open to any public organization and company to participate in, including non-Japanese. The program has two levels of participation, full membership (special supporting members) and associate membership (ordinary members). Full membership permits participation in all ISTEC activities. The special supporting members can send up to two researchers to the center for up to three years and have full access to the R&D results. Full membership requires an initial donation of about $1 million with annual charges of $120,000, while associate members cannot participate in the research center and have no access to the results. They can only participate in the symposia, subscribe to the publications and have access to the information and reference materials. The initial fee as well as the annual charges are $20,000. The total number of members, around 100, has been fairly stable through the years:

- 1988: 85 Japanese companies signed up as founding members.
- 1991: 112 members, of which 46 were special members (all Japanese companies) and 66 ordinary members (56 Japanese and 10 foreign companies).
- 1996: 103 members; of which 41 are special members (40 Japanese companies and one foreign company) and 62 ordinary members (52 Japanese and 10 foreign companies).

In 1996, the majority of full members have their main business in the fields of electric power (10 companies), consumer electronics (10), cable and wire (6), and iron and steel (4). The most important ordinary members are active in the businesses of construction and engineering (13 companies), electronic and telecommunication (8), banking and insurance (6), heavy machinery (4), gas (4), and metal and mining (4). There are also two research organizations participating as full members and six more as ordinary members in ISTEC. Foreign participation in ISTEC plays a minor role (about 10 percent of all members). However, the first foreign company recently became a full member (i.e. Du Pont de Nemours & Company). The ordinary foreign members are from the industrial sector (American Superconductor Corp.; Intermagnetics General Corp.; Hoechst AG; IBM Japan; Pirelli Cavi spa-Division Italia) and research organizations (Forschungszentrum Karlsruhe; Hewlett Packard Laboratories Japan; Korea Superconductor Research Center; CISE SPA).

The annual budget is in the range between $21 million and $32 million, as can be seen from Figure 13.1. In the beginning, funding was mainly conducted through initial donations of the special members. The share of the budget provided by the government, in this case NEDO, was increased every year from 14 percent in 1988 to 79 percent in 1995.

The increasing amount of money paid by the government (NEDO) might be seen as an answer to the request from industry to increase public budgets for research and development projects. In Japan, unlike other highly industrialized countries, public funding for research programs is generally rather small compared to the R&D budgets of industry. In total, MITI's budget for R&D in superconductor technology increased from $80.1 million in 1994 to $94.45 million in 1995 which translates into an increase of about 18 percent (ISTEC 1995: 71–2). Included in this amount is the budget for ISTEC which corresponded to 27 percent of the total in 1994.

In 1994, the ISTEC budget comprised the following items: 22 percent of the expenses were spent for personnel, 10 percent for material, 17 percent for capital investment, 8 percent for buildings and 43 percent for others (rent, electrical charges, etc.). An interesting fact is the relatively low percentage of

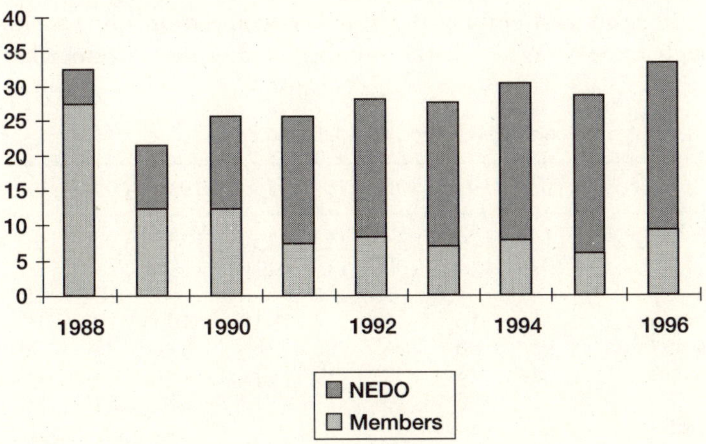

Figure 13.1 Budget of the superconductivity research laboratory (million US$)
Source: Internal ISTEC documents.

personnel spending which is usually in the range of 40–50 percent in this kind of institution. This is related to the fact that researchers from the special member-companies are mainly paid for by their companies. Hence, the budget for equipment and material is large, which allows the center to work with state-of-the-art equipment.

Structure and staff

The President of ISTEC is the Vice Chairman of *Keidanren*, the Japanese Federation of Economic Organizations. There are three Vice Presidents, currently the President of Toshiba Corp., the President of Fujitsu Ltd. and the Director General of SRL, one Senior Managing Director (a MITI official), two Managing Directors and forty-six other Directors. This structure reflects the links ISTEC established within Japanese scientific, economic and political circles.

ISTEC's associate SRL has one Director General and one Deputy Director General. The administrative work is done by the Planning and Management Division. There are eight Research Divisions which are divided into fundamental research on superconductors (Div. I to III), processing of superconductors (Div. IV, V and VII) and fundamental technology for applications (Div. VI and VIII). Division IX is responsible for public data base and there is one Development Promotion Division.

The staff of SRL comprises researchers, students and administrative staff. The researchers are further divided into division directors, employed, dispatched and visiting researchers. The staff are selected from different organizations and countries in order to bring together people with different knowledge and scientific backgrounds. The personnel structure is summarized in Table 13.6.

The division directors and employed researchers come from universities and other research centers. Whereas the number of division directors has

Table 13.6 Personnel of superconductivity research laboratory

	1989	*1990*	*1991*	*1992*	*1993*	*1994*	*1995*	*1996*
Total personnel	118	124	133	153	150	173	177	190
Researchers	108	114	108	114	116	129	128	126
Division directors	7	7	7	8	8	9	10	10
Employed researchers	5	16	5	6	10	19	10	19
Dispatched researchers	95	88	94	88	86	82	85	79
Visiting researchers	1	3	2	12	12	19	23	18
Students	0	0	10	11	17	21	23	27
Administrative staff members	10	10	15	28	17	23	26	27

Source: Internal ISTEC documents.

increased through the years, the number of employed researchers varies from year to year. This is rather uncommon because employed researchers are usually considered as the backbone of any research center as they collect knowledge and skill and train young researchers coming to the laboratory. This argument seems to play a minor role according to the data given in Table 13.6.

The special members of ISTEC send their own staff, dispatched researchers, to SRL for a period of two to three years. During their time in SRL, they are completely separated from their company and the research subject they are working on is defined within the divisions of SRL. In other words, the influence of the companies on the research topics is restricted, which is an important issue for carrying out independent basic research. However, the companies very often send untrained or even newly hired researchers to ISTEC. This means that some companies consider ISTEC as a center for education and training which was not the original goal as defined by MITI. On the other hand, in such a new area of materials it is difficult to find a large number of well trained people. The number of dispatched researchers declined from ninety-five in 1989 to seventy-nine in April 1996 reflecting the trend of member companies to reduce their activities within ISTEC.

There are also visiting researchers from non-Japanese research organizations. Most of these researchers are specialists, holding a doctorate degree and having experience in superconductivity research. Their number has increased constantly from one in 1989 to eighteen in April 1996 which shows the trend of internationalization within ISTEC. The total number of visiting researchers from 1988 to 1996 is sixty-five with fourteen from China, nine from Germany, eight from Korea, seven from the UK, seven from the United States, and twenty from other countries. Considering their origin by regions, Europeans (including Russians) comprise 45 percent, Asians 40 percent and North Americans 14 percent of the total number. The visiting researchers come to SRL through different fellowship systems; i.e. the ISTEC-fellowship, the NEDO-fellowship and the STA-fellowship which are awarded by the respective bodies. As of April 1996, there were four ISTEC fellows, four NEDO fellows and ten STA fellows. The fellowships are usually granted for between one to three years, but certain STA fellows come to SRL only for a short term of three months.

There are some unofficial links to universities which enable students to work in SRL. The number of students has increased constantly from ten in 1991 to thirty-seven in April 1996. They are enrolled in Master or PhD courses in engineering or science. Education and training does seem to play some role as there have been some thirty-seven doctorates awarded on research carried out at SRL since 1990. It should be mentioned that many dispatched researchers also obtained their PhD during their stay at SRL.

Merits and problems for the participating companies

Finally, the merits of participation for private partners in public R&D programs are summarized, and some critical comments regarding the situation in SRL will highlight a number of its weaknesses. Cooperation does have some advantages which are attractive for partners from industry as well as from university. Product development sometimes takes ten to twenty years from the first discovery or idea to the appearance of a product on the market. During that time, huge investments in private R&D departments are necessary, which means a high risk for companies. Especially at the beginning of product or process development, during the pre-competitive phase, involvement in public R&D projects is considered to be fruitful.

The benefits of cooperation are summarized in the following check list:

- Reduction of costs and risk.
- Critical mass.
- Time factor.
- Synergy in solving problems.
- New contacts between staff members.
- New ideas, knowledge and ability.

In ISTEC, reduction of costs and risk are realized by the possibility of member-companies to send their researchers to SRL to use the joint facility. In the set-up phase of ISTEC it was very important to obtain the basic funding for buying the very expensive equipment used in SRL. With eighty-five founding members the critical mass was given in 1988. The organization of workshops and seminars is possible for ISTEC but would be impossible for a single company.

The time factor and synergy effects are closely related to each other and will only be effective if the staff are flexible, trained to cooperate and have some experience in doing research. These benefits are restricted in cases where the companies send staff with little scientific background or even untrained freshmen. Then, a long period is necessary for the researchers to acclimatize to the work. Naturally, companies do not like to send their best researchers to such kinds of cooperation when strong in-house research in the same field is being carried out. The situation of ISTEC is no exception regarding this point. Training unskilled researchers takes up to one year and, therefore, the time remaining to obtain scientific and technical results is decreased drastically out of the total of two to three years that an average dispatched researcher stays at SRL. Despite this, the strategic argument of having the same access to the results for the different participating companies can be considered in the wider sense as synergy.

New contact between researchers with different scientific backgrounds coming from different laboratories obviously opens the way for creating new ideas, new knowledge and new abilities. Restrictions arise when patents are the result of a particular project. In such cases, the technical results are naturally confidential. Another point is the dispatched researchers' loyalty which is more dedicated to their own company in which their future career lies. From the viewpoint of foreign research fellows, communication is sometimes restricted due to the language barrier and culture gap. English, the *lingua franca* of science, is still less accepted in Japan than in other countries.

Summary

Several R&D programs were launched in the 1980s in the field of advanced materials by MITI, STA, the Ministry of Education and other government agencies. Advanced materials are a group of high value-added materials which have achieved an improved level of functional and structural characteristics. The four sub-groups, metals and alloys, ceramics, polymers and composites, find applications in various products and processes, especially in high-technology.

The most important sub-group of advanced materials are ceramics, the importance of which is based on the possibility of realizing new generations of microelectronics devices which is important to maintain the strength of the existing microelectronics industry. Furthermore, one sub-group of ceramics, high-temperature superconducting ceramics, shows even more possibilities for new high-technology products such as:

1 bulk materials for magnets and substrates in electronics;
2 tapes and wires for motors and generators; and
3 microelectronics devices.

Japanese industry and government research centers have been involved since the early 1970s in R&D programs for low-temperature superconductors as well as for ceramics (Lastres 1994: 103–5). Therefore, the Japanese response to the discovery of high-temperature superconducting ceramics was a quick one. Several programs were launched by MITI, STA and the Ministry of Education in 1987. Those of MITI and STA have some distinctive characteristics, like cooperation with industry, long duration periods (up to ten years), and gathering and diffusion of information among the participants. Furthermore, in some of the programs there was an emphasis on the promotion of creative basic research according to the new R&D policy mentioned above.

ISTEC is one of these MITI projects in the field of superconductivity with a rather broad target, i.e. all aspects of superconducting materials, from theory to process development, are investigated. The trend of MITI to promote internationalization is reflected in the opportunity for non-Japanese partners to participate in ISTEC, the international symposia organized by ISTEC, and ISTEC's invitation of foreign research fellows.

The research itself is carried out in ISTEC's affiliated SRL, which is independent from ISTEC as well as from the member companies regarding decisions of the actual research topics being investigated. The staff of SRL come from the Japanese member companies, universities and non-Japanese institutions. The very favorable financial situation allows SRL to purchase state-of-the-art equipment and to use high quality materials which are necessary to work in high-technology fields such as superconductivity. A drawback concerning the staff at SRL, however, is the relatively small number of permanent researchers on the one side compared with the large number of inexperienced dispatched researchers as well as the increasing number of students on the other side.

In 1996, the first products making use of high-temperature superconducting materials are ready to enter the market. However, only a few of the expected new developments have been realized so far. There is still a need for further public funding of R&D projects in high-temperature superconductivity. In Japan as well as in the US, superconducting materials are considered as an important sub-group of advanced materials and the government budgets for these research fields are still considerable.

The data presented in this report shows that the Japanese R&D activities in advanced materials have been quite intensive in the past fifteen years. Several programs for development as well as for basic studies of new materials were launched by different ministries and government agencies. They reflect a Japanese policy to develop new products and markets and also to extend the trend to include basic science in development and engineering concentrated research programs. Since 1986, the year of the discovery of high-temperature superconducting materials, Japanese R&D programs focused stronger on superconductivity research. Although the present number of products using superconducting materials is rather small world-wide, the knowledge and experience of fabrication is at a high level in Japan. This knowledge together with the financial backbone of the R&D programs should be a good basis for the future production of advanced materials including superconducting material products with the label 'Made in Japan.'

NOTES

1 The data for this case study was obtained in June 1996 from ISTEC.

REFERENCES

Abraham, T. (1995) The US Advanced Ceramic Industry. In *Materials Technology*, vol. 10, nos. 3/4, pp. 68–72.

Hane, G. (1991) *Research and Development Consortia and Cooperative Relationships in Japan's Superconductivity Industry.* Tokyo: National Institute of Science and Technology Policy (NISTEP).

ISTEC (1995) Budgets for Major Superconductivity Related Projects. In *ISTEC Journal*, no. 8, p. 71.

JETRO (1990) *Your Market in Japan: New Materials (2nd. ed.)*, March 1990. Tokyo: Japan External Trade Organization Tokyo (JETRO).

Kenward, M. (1996) Superconductors Get Ready to Transform Industry. In *Physics World*, June, pp. 29–31.

Lastres, H.M.M. (1994) *The Advanced Materials Revolution and the Japanese System of Innovation*. New York: St Martin's Press.

Lubkin, G.B. (1996) Power Applications of High-Temperature Superconductors. In *Physics Today,* March, pp. 48–51.

MK Associates (1995) *An Overview of US Research and Development in Superconductivity, Prepared for JETRO New York*, February 1990. Washington, DC: MK Associates.

Narula, C.K., J.E. Allison, D.R. Bauer and H.S. Gandhi (1996) Materials Chemistry Issues Related to Advanced Materials Application in the Automotive Industry. In *Chemistry of Materials*, vol. 8, pp. 984–1003.

Simon, R. (1995) NMR: The First Big Market for HTS Technology. In *Superconducting Industry,* Summer 1995, pp. 18–23.

INDEX